Watershed Dynamics

NIPA® GENX ELECTRONIC RESOURCES & SOLUTIONS P. LTD.
New Delhi-110 034

About the Author

Dr. Vartika Singh is an esteemed researcher and academician with a profound background in geology and hydrogeology. She holds a Ph.D. and M.Phil. in hydrogeology from Vikram University and a Master's degree in applied geology from Kurukshetra University. With 15 years of research experience at the Defence Terrain Research Laboratory (DTRL), DRDO, Delhi, and Amity University Noida, Dr. Singh has significantly advanced her field. Her research expertise spans hydrogeology, remote sensing, environmental geology, groundwater, watershed management, and artificial intelligence. Dr. Singh's substantial research achievements are noteworthy. She has published over 60 scientific papers, holds three patents, and has authored three books and 20 book chapters, demonstrating her dedication to scholarly pursuits. As an Assistant Professor at the Amity Institute of Global Warming and Ecological Studies (AIGWES) at Amity University Noida, she contributes to shaping the academic landscape and mentors several research students, showcasing her commitment to guiding the next generation of researchers. Her involvement in the editorial boards of several national and international journals underscores her active participation in the peer-review process and her role in disseminating knowledge within the academic community. Dr. Vartika Singh's impressive research experience, prolific publication record, and active engagement in academia underscore her expertise and significant contributions to the fields of hydrogeology, remote sensing, and environmental geology.

Watershed Dynamics
Strategies and Practices for Sustainable Development

Vartika Singh
Assistant Professor
Amity Institute of Global Warming and Ecological Studies (AIGWES)
Amity University
Noida, Uttar Pradesh

NIPA® GENX ELECTRONIC RESOURCES & SOLUTIONS P. LTD.
New Delhi-110 034

NIPA® GENX ELECTRONIC RESOURCES & SOLUTIONS P. LTD.

101,103, Vikas Surya Plaza, CU Block
L.S.C. Market, Pitam Pura, New Delhi-110 034
Ph : +91-11-43860225, Mob.: +91 9717133558, 9540816132
E-mail: newindiapublishingagency@gmail.com
Website: www.nipaersources.com

Print ISBN: 978-93-58874-41-9
ebook ISBN: 978-93-58872-10-1

Composed and Designed by NIPA®.

Dedicated to
The Sweet Memory of my Beloved Husband

Late Shri Ekank Rana
Who was a source of Inspiration to me.

Preface

Watersheds are important landscapes that function as natural units for managing water, soil, and other natural resources. These systems' dynamics are complicated, including a variety of environmental, socioeconomic, and technical aspects. As mankind faces the multifaceted issues of climate change, population increase, and sustainable development, the need for good watershed management has never been greater. "Watershed Dynamics: Strategies and Practices for Sustainable Development" seeks to provide a thorough guide to understanding, managing, and preserving these critical ecosystems. This book is organised to cover a wide variety of issues relevant to integrated watershed management. It begins with an overview of the subject, setting the groundwork by investigating historical circumstances and changing methods. We look at the concepts and objectives that motivate

The socioeconomic situations of populations in watershed regions are critical in developing management measures. We give a detailed look at these communities' demographic characteristics, livelihood patterns, and economic activities. Social institutions and governance systems are examined to better understand their role in promoting sustainable watershed management. The book also discusses community vulnerability and resilience to environmental change, providing insights on how to develop better, more adaptable communities. The value of participatory methodologies, multi-stakeholder engagement, and partnership formation is emphasised. Practical information is offered for planning and implementing watershed development initiatives, as well as frameworks for monitoring, evaluation, and adaptive management approaches. Case studies and success stories are provided to illustrate practical applications and lessons learned. The advent of remote sensing technology revolutionised watershed management. This book examines remote sensing technologies and techniques, as well as their applications and interactions with other technologies such as GIS. The interdisciplinary approach adopted in this book reflects the complexities of watershed management and underlines the significance of collaboration in achieving sustainable development goals. We hope that this book will inspire and motivate you in your efforts to better understand and manage watersheds, leading in a more sustainable and harmonious balance between human activity and nature.

I am deeply grateful for the support and cooperation of my family, friends, and colleagues throughout the process of creating this volume. Their sincere assistance has been invaluable. I would like to extend special thanks to Shri J.C. Kala, IFS (Retd), Honorary Adviser at the Amity Institute of Global Warming and Ecological Studies, Amity University, Noida, whose guidance and expertise greatly contributed to the development of this work. I am also profoundly thankful to Dr. V. K. Panchal, Director (Retd) of DTRL, DRDO, New Delhi, for their contributions and support. Additionally, I express my heartfelt gratitude to the late Mr. Ekank Rana, Director of Eroth Tech (Mohsri Care Pvt. Ltd.), for their encouragement and blessing in the creation of this volume. Their collective efforts and expertise have played a significant role in making this project possible, and I am truly grateful for their contributions.

I am sincerely thankful to M/s NIPA® Publisher for Publishing this book.

Noida **Vartika Singh**

Contents

1

Integrated Watershed Management

1.1 Introduction to Integrated Watershed Management

Definition and Scope of Integrated Watershed Management: Integrated Watershed Management (IWM) refers to a comprehensive approach to managing the resources and ecological processes within a watershed area in a coordinated and integrated manner. Integrated watershed management represents a holistic approach to managing natural resources within a watershed, with the aim of achieving sustainable development while maintaining ecosystem health and resilience. It involves collaboration among various stakeholders and integrates scientific knowledge with traditional wisdom to address complex environmental challenges. This section aims to provide a clear understanding of what integrated watershed management entails, including its scope and key components.

a. **Definition of Integrated Watershed Management:** Integrated Watershed Management is a holistic and multi-disciplinary approach that considers the entire watershed as a unit for planning and managing natural resources. It involves the coordinated management of land, water, vegetation, and other resources within a watershed to achieve sustainable development goals while preserving ecological integrity. Here are several definitions from various sources:

United Nations Environment Programme (UNEP)

"Integrated watershed management is a process which promotes the coordinated development and management of water, land, and related resources across sectors within a given river basin, in order to maximize the economic and social benefits derived from water resources in an equitable manner while preserving and, where necessary, restoring freshwater ecosystems."

World Bank

"Integrated Watershed Management is an approach that promotes the sustainable management of natural resources to achieve environmental, social, and economic objectives within a watershed context. It involves coordinated planning, implementation, and monitoring of activities that address the interlinked challenges of water scarcity, soil erosion, biodiversity loss, and poverty."

Food and Agriculture Organization (FAO)

"Integrated Watershed Management is a process that involves the development and implementation of strategies to sustainably manage land, water, and related resources within a watershed. It emphasizes the importance of stakeholder participation, multi-sectoral collaboration, and adaptive management approaches to address the complex and interconnected challenges facing watersheds."

Environmental Protection Agency (EPA), United States

"Integrated Watershed Management is a comprehensive approach to managing the natural resources within a watershed in a coordinated manner. It involves the integration of scientific knowledge, stakeholder input, and management practices to achieve environmental protection, sustainable development, and community resilience."

International Water Management Institute (IWMI)

"Integrated Watershed Management is an approach that seeks to balance competing water uses and ecosystem needs within a watershed by considering the interactions between hydrology, ecology, and socio-economic factors. It aims to promote the sustainable use of water resources while safeguarding ecosystem health and enhancing community livelihoods."

b. **Scope in Terms of Geographical Coverage**: Integrated watershed management typically encompasses a defined geographical area delineated by natural boundaries such as ridgelines, hills, or mountains, where all surface water drains to a common point, such as a river, lake, or ocean. The scope of IWM may vary depending on factors such as watershed size, topography, and hydrological characteristics.

c. **Ecosystem Components:** IWM considers various ecosystem components within the watershed, including soil, water, vegetation, and wildlife. It recognizes the interconnectedness of these components and aims to maintain or restore their ecological balance through sustainable management practices.

d. **Management Objectives:** The primary objectives of integrated watershed management include but are not limited to:

- Conservation and sustainable use of natural resources
- Protection of water quality and quantity
- Soil conservation and erosion control

- Biodiversity conservation and habitat restoration
- Climate change adaptation and resilience building
- Socio-economic development and livelihood improvement

e. **Stakeholders Involved:** Integrated watershed management involves a wide range of stakeholders, including government agencies, local communities, non-governmental organizations (NGOs), researchers, private sector entities, and other relevant actors. Stakeholder participation is essential for effective planning, implementation, and monitoring of watershed management activities, as it ensures that diverse perspectives and interests are taken into account.

Significance of Integrated Approaches in Environmental Conservation

The significance of integrated approaches in environmental conservation, particularly in the context of India, stems from the complex interplay of various environmental components within watersheds and the need for comprehensive strategies to address interconnected challenges. Integrated approaches are essential for effective environmental conservation in India due to the interconnectedness of environmental components within watersheds and the need for comprehensive strategies to address complex challenges. By adopting holistic management approaches, India can maximize conservation impact, promote sustainable resource management, engage local communities, and enhance resilience to climate change. Here's how integrated approaches are crucial for effective environmental conservation in India:

a. **Interconnected Environmental Components:** In India, watersheds are not isolated entities but rather intricate systems where various environmental components interact. These components include soil, water, vegetation, wildlife, and even human activities. Integrated approaches recognize these interconnections and aim to manage them holistically rather than addressing individual issues in isolation.

b. **Addressing Complex Environmental Challenges:** India faces a myriad of environmental challenges, including deforestation, soil erosion, water pollution, loss of biodiversity, and climate change impacts. These challenges are often intertwined and cannot be effectively tackled in isolation. Integrated approaches acknowledge the complexity of these issues and seek to address them comprehensively within the context of watersheds.

c. **Maximizing Conservation Impact:** By adopting integrated approaches, India can maximize the impact of conservation efforts. Rather than implementing fragmented initiatives that target specific aspects of environmental degradation, integrated management strategies consider the entire ecosystem within watersheds. This approach allows for synergies between different conservation measures, leading to more efficient and effective outcomes.

d. **Sustainable Resource Management:** Integrated approaches promote sustainable management of natural resources within watersheds. By considering the interlinkages between soil, water, vegetation, and wildlife, these approaches aim to maintain ecological balance while meeting the socio-economic needs of local communities. This ensures the long-term viability of natural resources and livelihoods dependent on them.

e. **Community Engagement and Participation**: Integrated approaches emphasize the involvement of local communities in conservation efforts. In India, where many watersheds are inhabited by rural populations, community participation is crucial for the success of environmental conservation initiatives. Integrated approaches empower communities to take ownership of conservation activities, fostering a sense of responsibility and stewardship towards their natural environment.

f. **Resilience to Climate Change:** With climate change posing significant challenges to India's ecosystems, integrated approaches play a vital role in enhancing resilience. By considering climate change impacts in watershed management planning and implementation, these approaches help communities adapt to changing environmental conditions, mitigate risks, and build resilience to extreme weather events.

Overview of Challenges Addressed by Integrated Watershed Management

In the Indian context, Integrated Watershed Management (IWM) is deployed to address a multitude of challenges that affect the sustainability of ecosystems and livelihoods. Here's an overview of the key challenges tackled by IWM, supported by an Indian case study:

a. **Soil Erosion:** Soil erosion is a significant challenge in India, especially in hilly and semi-arid regions where agricultural practices and deforestation exacerbate soil degradation. IWM employs techniques such as contour bunding, terracing, and afforestation to mitigate erosion and conserve soil fertility.

- **Indian Case Study:** The Ralegan Siddhi watershed in Maharashtra is a notable example where community-led watershed management initiatives, including afforestation, soil conservation, and rainwater harvesting, have successfully reduced soil erosion and improved agricultural productivity.

b. **Water Scarcity:** India faces water scarcity due to overexploitation of groundwater, inefficient water management practices, and erratic rainfall patterns exacerbated by climate change. IWM focuses on water harvesting, groundwater recharge, and efficient irrigation techniques to alleviate water scarcity and enhance water availability for both agricultural and domestic purposes.

   - **Indian Case Study**: The Tarun Bharat Sangh's watershed management efforts in Alwar district, Rajasthan, have transformed barren landscapes into water-rich areas through rainwater harvesting structures, check dams, and community-managed ponds, thereby mitigating water scarcity and improving agricultural productivity.

c. **Biodiversity Loss**: Habitat destruction, fragmentation, and pollution have led to biodiversity loss in India, threatening ecosystems and wildlife. IWM emphasizes habitat restoration, conservation of native species, and the protection of ecological corridors to safeguard biodiversity within watersheds.

   - **Indian Case Study**: The Western Ghats, a biodiversity hotspot in India, faces threats from deforestation and land-use change. Integrated watershed management initiatives in Kerala and Karnataka focus on reforestation, protection of wildlife habitats, and community-based conservation efforts to preserve biodiversity in these ecologically sensitive regions.

d. **Socio-economic Vulnerabilities**: Rural communities in India are vulnerable to poverty, food insecurity, and livelihood challenges exacerbated by environmental degradation. IWM integrates livelihood enhancement programs, skill development, and social safety nets to improve the socio-economic resilience of communities dependent on watershed resources.

   - **Indian Case Study**: The Watershed Development Program implemented in tribal regions of Gujarat has not only improved soil and water conservation but also empowered local communities through capacity building, alternative livelihood generation, and participatory decision-making, thereby reducing socio-economic vulnerabilities.

Table: Table provides an overview of the diverse challenges addressed by integrated watershed management in India. By adopting a holistic and participatory approach, integrated watershed management aims to address environmental, social, and economic challenges to promote sustainable development and resilience in watersheds.

S. No.	Challenge	Description
1.	Soil Erosion	Soil erosion is a major issue in many watersheds, leading to loss of fertile soil and reduced agricultural productivity. Integrated watershed management aims to implement soil conservation measures such as contour farming, terracing, and afforestation to mitigate erosion.
2.	Deforestation	Deforestation contributes to loss of vegetation cover, habitat fragmentation, and reduced ecosystem services. Integrated watershed management focuses on reforestation, afforestation, and sustainable forest management practices to restore degraded landscapes and conserve biodiversity.
3.	Declining Water Availability	Water scarcity is a significant challenge in many regions, exacerbated by over-extraction, pollution, and climate change. Integrated watershed management aims to enhance water availability through measures such as rainwater harvesting, groundwater recharge, and efficient irrigation practices.
4.	Floods and Natural Disasters	Floods and other natural disasters pose risks to communities and infrastructure, leading to loss of lives and property. Integrated watershed management includes flood risk assessment, early warning systems, and nature-based solutions such as wetland restoration and floodplain zoning to mitigate disaster risks.
5.	Degradation of Water Quality	Pollution from agricultural runoff, industrial discharge, and untreated sewage contaminates water bodies, affecting human health and aquatic ecosystems. Integrated watershed management implements measures such as riparian buffers, vegetative filter strips, and wastewater treatment to improve water quality.
6.	Loss of Biodiversity	Habitat loss, fragmentation, and invasive species threaten biodiversity in watersheds, leading to loss of species and ecosystem services. Integrated watershed management focuses on conserving and restoring habitats, establishing ecological corridors, and controlling invasive species to protect biodiversity.
7.	Socioeconomic Inequities	Socioeconomic disparities contribute to unequal access to resources and opportunities, exacerbating vulnerability to environmental risks. Integrated watershed management promotes participatory approaches, equitable resource allocation, and livelihood diversification to address socioeconomic inequities and enhance community resilience.

These challenges comprehensively, Integrated Watershed Management in India aims to enhance environmental sustainability, improve socio-economic well-being, and build resilience to climate change, thereby ensuring the long-term viability of ecosystems and livelihoods dependent on watershed resources.

1.2 Understanding the Concept of Watershed Management

- **Characteristics of a Watershed**: A watershed, also known as a drainage basin or catchment area, is a fundamental geographic unit that plays a crucial role in the management and distribution of water resources. The characteristics of a watershed, including its natural boundaries, drainage patterns, and common outlet, are essential for understanding the dynamics of water flow, distribution, and management within a geographical region. Watersheds play a critical role in managing water resources and are integral to maintaining ecological balance and supporting human livelihoods. This subsection aims to provide a detailed understanding of the key characteristics of watersheds and their significance in water resource management:

a. **Definition of a Watershed**: A watershed is defined as an area of land bounded by natural topographic features, such as ridges, hills, or mountains, that separates it from adjacent areas. It is a geographical region where all surface water drains to a common point, such as a river, lake, or ocean. The boundary of a watershed is determined by the highest elevations, also known as the watershed divide or drainage divide, which directs the flow of water towards the lowest points.

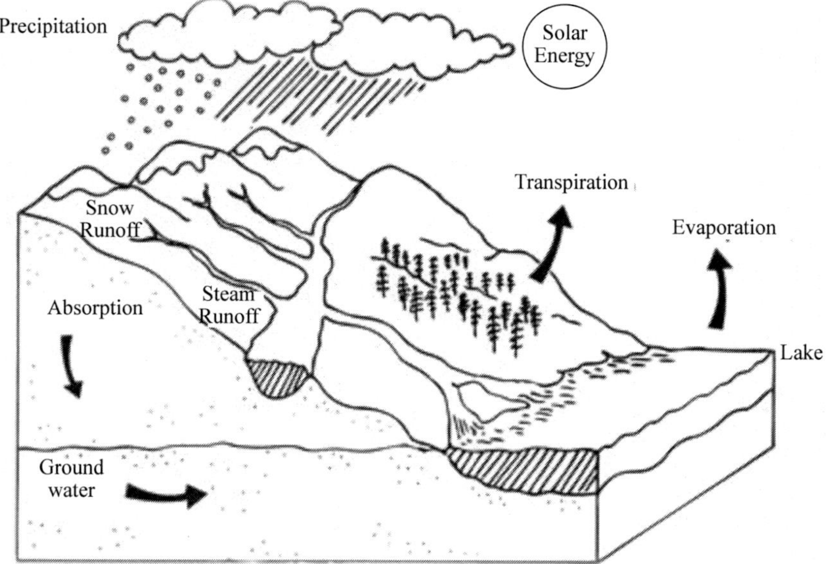

Figure 1. Watershed Illustration

b. **Natural Boundaries**: Watersheds are delineated by natural boundaries that define their extent and shape. These boundaries are typically defined by ridgelines or high points along the terrain, where water flows down-slope in multiple directions. The boundary of a watershed acts as a physical barrier that separates it from neighboring watersheds, influencing the direction and path of water flow within the basin.

c. **Drainage Patterns:** Watersheds exhibit diverse drainage patterns characterized by the arrangement of streams, rivers, and tributaries within the basin. Common drainage patterns include dendritic, parallel, radial, and trellis patterns, which are determined by factors such as topography, geology, and climate. The drainage pattern of a watershed influences the flow of water, sediment transport, and nutrient cycling within the basin.

d. **Concept of a Common Outlet**: One of the defining features of a watershed is the concept of a common outlet, where all surface water within the basin ultimately converges and exits the watershed. This common outlet could be a river mouth, estuary, lake, or ocean depending on the geographic location of the watershed. The common outlet serves as the endpoint of the hydrological cycle within the watershed, where water is discharged into larger water bodies.

e. **Importance in Water Resource Management:** Watersheds are fundamental units for managing water resources as they serve as the primary source of surface water and groundwater for human activities, agriculture, industry, and ecosystems. Understanding the characteristics of watersheds is essential for assessing water availability, quality, and distribution within a given region. Watershed management involves implementing strategies to sustainably manage water resources, conserve natural habitats, and protect ecosystems within the basin.

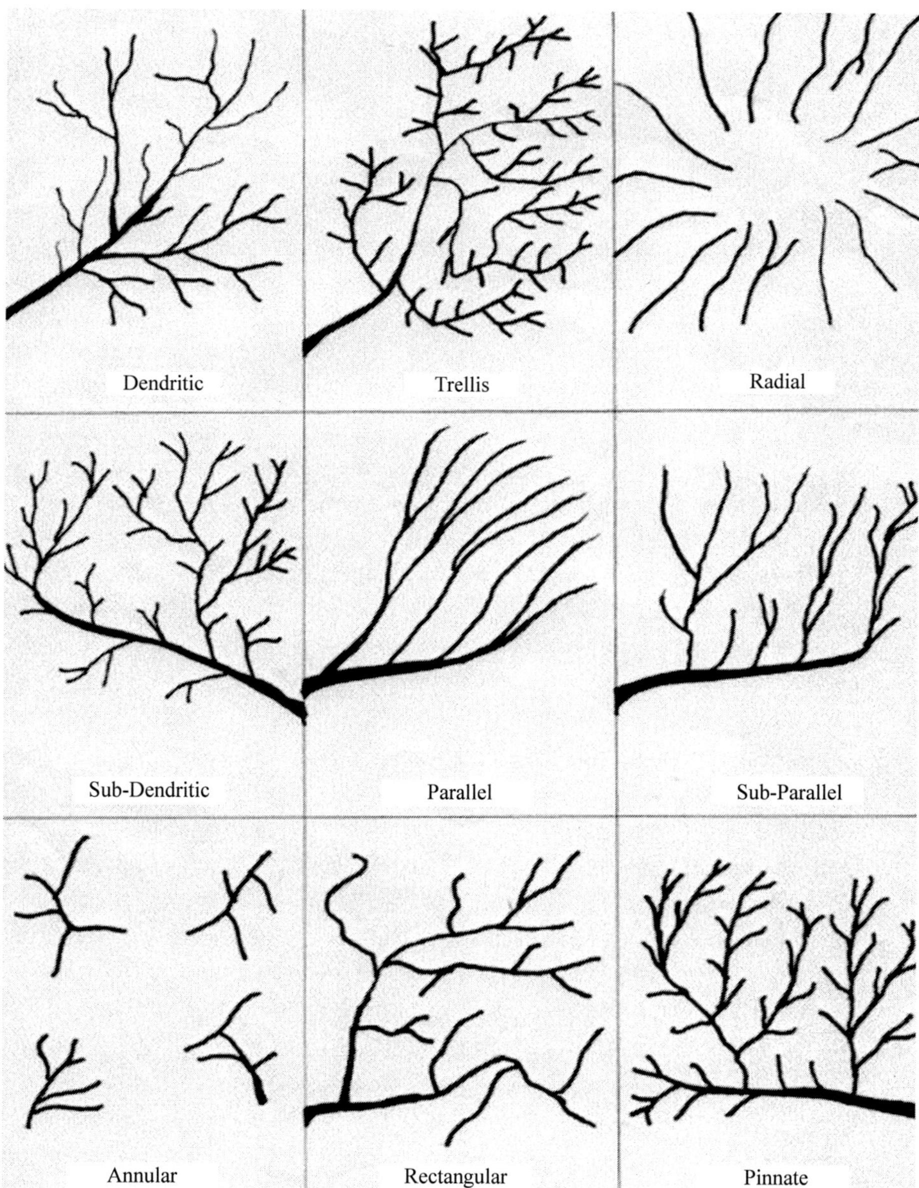

Figure 2. Different Drainage Patterns

Holistic Approach to Watershed Management

A holistic approach to watershed management recognizes the interconnectedness of various components within a watershed ecosystem and emphasizes the need to address multiple factors simultaneously to achieve sustainable outcomes. A holistic approach to watershed management recognizes the interdependence of soil, water, vegetation, and wildlife within a watershed ecosystem and seeks to address multiple components simultaneously through integrated management strategies. By adopting holistic approaches, watershed managers can maximize ecosystem benefits, enhance resilience, and promote long-term sustainability for both natural ecosystems and human communities. This subsection explores the principles and benefits of adopting a holistic approach in watershed management:

a. **Interconnectedness of Soil, Water, Vegetation, and Wildlife**: A watershed is a complex ecological system where soil, water, vegetation, and wildlife are intricately interconnected. Changes in one component can have cascading effects on the entire ecosystem. For example, deforestation can lead to soil erosion, which affects water quality and aquatic habitats, ultimately impacting wildlife populations. A holistic approach acknowledges these interdependencies and seeks to manage the entire watershed as a unified system.

b. **Integrated Management Strategies**: Holistic watershed management involves the integration of various management strategies that address multiple components simultaneously. Rather than focusing on individual aspects in isolation, such as soil conservation or water quality improvement, holistic approaches consider the interactions and feedbacks between different components. For instance, restoring riparian vegetation not only stabilizes stream banks but also improves water quality and provides habitat for wildlife.

c. **Synergistic Benefits:** By addressing multiple components of the watershed ecosystem, holistic management approaches can generate synergistic benefits that enhance overall ecosystem health and resilience. For example, reforestation efforts not only sequester carbon and mitigate climate change but also regulate water flow, prevent soil erosion, and support biodiversity conservation. These synergies amplify the effectiveness of management interventions and contribute to more sustainable outcomes.

d. **Comprehensive Planning and Decision-Making**: Holistic watershed management involves comprehensive planning and decision-making processes that take into account the diverse needs and objectives of stakeholders, including ecological, social, and economic considerations. This often requires interdisciplinary collaboration among scientists, policymakers, local communities, and other stakeholders to develop integrated management plans that balance competing interests and priorities.

e. **Long-Term Sustainability**: By considering the interconnectedness of soil, water, vegetation, and wildlife, holistic watershed management aims to achieve long-term sustainability by preserving ecosystem integrity and supporting human well-being. Sustainable management practices, such as agroforestry, wetland restoration, and sustainable agriculture, are integral components of holistic approaches that promote resilience and adaptability to changing environmental conditions.

Components of Watershed Management

Watershed management involves the integrated management of various components within a watershed ecosystem to promote ecosystem health, resilience, and sustainability. By addressing these key components of watershed management through integrated and sustainable approaches, stakeholders can promote ecosystem health, resilience, and sustainability, thereby ensuring the long-term viability of watershed ecosystems and the services they provide to both nature and society. This section breaks down the key components of watershed management and highlights their roles in maintaining ecosystem functionality:

a. **Soil Management: Role:** Soil is a fundamental component of watershed ecosystems, playing crucial roles in nutrient cycling, water retention, and supporting plant growth. Healthy soils are essential for maintaining soil fertility, preventing erosion, and sustaining agricultural productivity.

Management Strategies

- **Soil Conservation Practices:** Implementing erosion control measures such as contour plowing, terracing, and cover cropping to reduce soil erosion and sedimentation in water bodies.
- **Soil Health Improvement:** Adopting practices such as organic farming, composting, and crop rotation to enhance soil fertility, structure, and resilience to degradation.

- **Land Use Planning:** Implementing land use zoning and conservation measures to minimize soil disturbance and protect critical soil resources from degradation.

b. **Water Management: Role:** Water is a vital resource within watersheds, supporting various ecosystem functions, human activities, and livelihoods. Effective water management is essential for ensuring water availability, quality, and sustainability within the watershed.

Management Strategies

- **Water Harvesting Techniques:** Constructing structures such as check dams, contour trenches, and rooftop rainwater harvesting systems to capture and store rainwater for irrigation, domestic use, and groundwater recharge.
- **Watershed Restoration:** Restoring degraded water bodies, such as rivers, wetlands, and ponds, through dredging, re-vegetation, and aquatic habitat enhancement to improve water quality and aquatic biodiversity.
- **Sustainable Irrigation Practices**: Promoting water-efficient irrigation techniques such as drip irrigation, sprinkler irrigation, and laser leveling to minimize water wastage and enhance agricultural water productivity.

c. **Vegetation Management: Role**: Vegetation, including forests, grasslands, and riparian vegetation, plays critical roles in watershed ecosystems, such as soil stabilization, water regulation, carbon sequestration, and habitat provision for wildlife.

Management Strategies

- **Reforestation Efforts**: Planting native tree species and restoring degraded forest areas to enhance biodiversity, soil stability, and ecosystem services such as carbon sequestration and water regulation.
- **Agroforestry Practices:** Integrating trees and shrubs into agricultural landscapes to improve soil fertility, provide shade and windbreaks, and diversify crop production while promoting ecological resilience.
- **Riparian Buffer Zones:** Establishing vegetative buffers along streams and water bodies to reduce sedimentation, filter pollutants, and provide habitat for aquatic species, thereby improving water quality and riparian ecosystem health.

d. **Wildlife Management**: **Role:** Wildlife, including flora and fauna, contributes to the ecological diversity, resilience, and functionality of watershed ecosystems. Conservation of wildlife habitats and species is essential for maintaining ecological balance and supporting ecosystem services.

Management Strategies

- **Habitat Restoration Projects:** Restoring degraded habitats, such as wetlands, grasslands, and forests, to provide suitable habitats for wildlife species and enhance biodiversity within the watershed.
- **Protected Area Management:** Establishing and managing protected areas, wildlife reserves, and conservation corridors to safeguard critical habitats and protect endangered species from habitat loss and fragmentation.
- **Sustainable Land Use Practices:** Promoting sustainable land use practices such as wildlife-friendly farming, agro-ecological zoning, and land set-asides to minimize habitat destruction and support wildlife conservation efforts.

1.3 Historical Context and Evolution of Integrated Approaches

Traditional Practices and Wisdom in Watershed Management

Traditional knowledge and practices related to watershed management have been developed over centuries by indigenous communities around the world, including in India. This section delves into the historical context of traditional practices and their continued relevance in modern-day watershed management approaches:

a. **Historical Context:** Traditional practices in watershed management have roots in indigenous cultures and local ecological knowledge systems that have evolved over generations. In India, traditional communities have developed sophisticated techniques for soil and water conservation, influenced by factors such as climate, topography, and socio-cultural norms. These practices often stem from a deep understanding of local ecosystems and their dynamics. In traditional practices and wisdom in watershed management embody centuries-old knowledge systems that have sustained ecosystems and communities for generations. Recognizing the value of indigenous knowledge and practices is essential for fostering resilience, promoting sustainability, and preserving cultural heritage within watersheds in India and beyond.One notable Indian case study that exemplifies traditional practices in watershed management is the "Johads of Rajasthan." Johads are traditional rainwater harvesting structures built by communities in the arid and semi-arid regions of Rajasthan, primarily in the Alwar district. These structures have been integral to water management and agricultural practices in the region for centuries.

Case Study: Johads of Rajasthan

Historical Context

The practice of constructing johads dates back centuries, with historical records indicating their existence during the medieval era. These structures were traditionally built by local communities, including farmers, herders, and village councils, to capture and store rainwater during the monsoon season for agricultural and domestic use.

Key Features

- **Design and Construction:** Johads are earthen check dams constructed across seasonal streams or in natural depressions to capture rainwater runoff. They are typically made of local materials such as stone, mud, and clay, with a sloping embankment on the upstream side to channel water into the reservoir.
- **Functionality:** During the monsoon season, johads collect rainwater, which percolates into the ground, replenishing groundwater aquifers and recharging wells. The stored water is used for irrigation, livestock watering, and domestic purposes during the dry season when surface water sources are scarce.
- **Community Management**: Johads are managed and maintained by local communities through collective efforts. Traditional governance systems, such as village councils (panchayats) and water user associations, oversee the equitable distribution of water and resolve conflicts over water rights and usage.

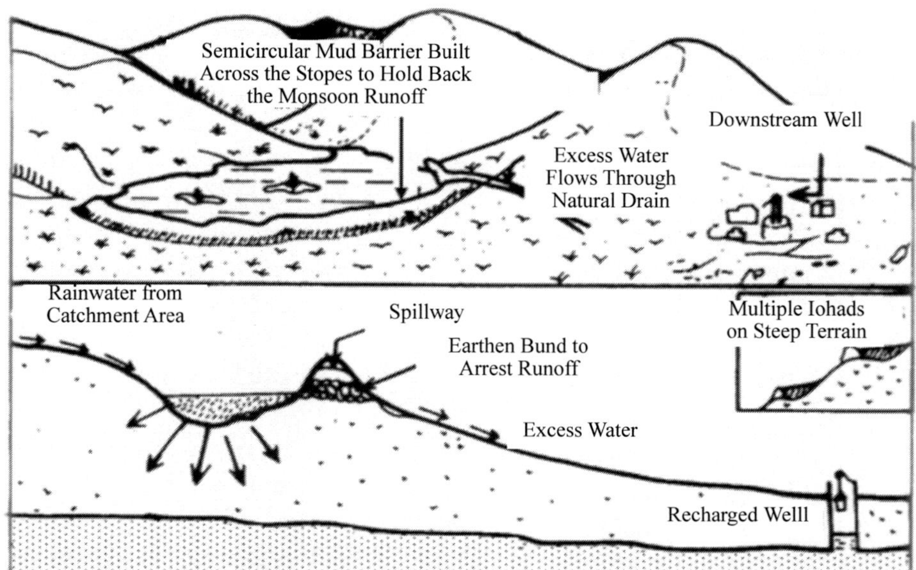

Figure 3. Johads

Relevance and Impact

The johads of Rajasthan have had significant socio-economic and environmental impacts on local communities and ecosystems:

Water Security: Johads have contributed to water security and resilience in water-stressed regions by providing reliable access to water for agriculture, livestock rearing, and household needs throughout the year.

Soil Conservation: The check dams built by johads help in soil conservation by reducing soil erosion and sedimentation, thereby improving soil fertility and agricultural productivity.

Biodiversity Conservation: The presence of perennial water sources created by johads supports diverse flora and fauna, including native plant species, migratory birds, and small mammals, enhancing biodiversity in the region.

Challenges and Opportunities

While johads have been successful in addressing water scarcity and supporting rural livelihoods, they face challenges such as silting, encroachment, and insufficient maintenance. Revitalizing and preserving traditional water harvesting practices like johads require concerted efforts to involve local communities, leverage modern technology for restoration and maintenance, and integrate indigenous knowledge with scientific approaches in watershed management.

The johads of Rajasthan exemplify the effectiveness and sustainability of traditional practices in watershed management, showcasing how indigenous knowledge and community-driven initiatives can address water scarcity and promote resilience in arid and semi-arid regions.

b. **Conserving Soil and Water Resources**: Traditional practices focus on conserving soil and water resources to ensure agricultural productivity, food security, and ecological resilience. Techniques such as contour farming, bunding, terracing, and agroforestry are commonly employed to prevent soil erosion, enhance soil fertility, and recharge groundwater aquifers. Indigenous water harvesting structures, including tanks, ponds, and stepwells, are designed to capture and store rainwater for dry-season irrigation and domestic use.

c. **Community-Based Management**: Traditional watershed management practices are often community-driven, with local communities actively involved in decision-making, implementation, and maintenance of conservation measures. Social norms, traditional governance systems, and customary laws play vital roles in regulating access to and use of natural resources within watersheds, fostering a sense of collective responsibility and stewardship towards the environment.

d. **Relevance in Modern-Day Management Approaches**: Despite advancements in technology and changes in socio-economic conditions, traditional practices in watershed management remain relevant in modern-day approaches. Many traditional techniques have been adapted and integrated into contemporary conservation strategies, owing to their effectiveness, cost-efficiency, and sustainability. Indigenous knowledge systems also offer valuable insights into ecosystem dynamics, resilience, and adaptation strategies in the face of climate change.

e. **Cultural and Ethical Dimensions:** Traditional watershed management practices are often imbued with cultural and ethical dimensions, reflecting spiritual beliefs, cultural values, and reciprocal relationships with nature. Practices such as sacred groves, community rituals, and customary water-sharing arrangements contribute to the conservation of biodiversity, preservation of sacred landscapes, and maintenance of cultural heritage within watersheds.

f. **Challenges and Opportunities:** While traditional practices offer valuable lessons for sustainable watershed management, they also face challenges such as socio-economic changes, demographic shifts, and environmental degradation. Revitalizing and preserving traditional knowledge systems require efforts to recognize, document, and promote indigenous practices, empower local communities, and integrate traditional wisdom with scientific approaches in watershed management planning and implementation.

Development of Modern Scientific Approaches

The evolution of watershed management from traditional practices to modern scientific approaches reflects a transition towards more systematic, evidence-based, and interdisciplinary approaches. The development of modern scientific approaches in watershed management represents a synthesis of traditional knowledge, scientific research, technological innovation, and interdisciplinary collaboration. By integrating indigenous wisdom with evidence-based practices, contemporary watershed management aims to address complex environmental challenges and promote sustainable resource management for present and future generations. This section traces the key milestones in this development and highlights the role of scientific research, technological advancements, and interdisciplinary collaboration in shaping contemporary management strategies:

a. **Historical Roots in Traditional Practices:** Watershed management has ancient roots in indigenous knowledge and traditional practices, such as soil conservation, water harvesting, and agroforestry, developed by communities over centuries to sustainably manage natural resources.

   Traditional approaches focused on localized solutions, community participation, and adaptive management strategies based on observation, trial and error, and cultural wisdom.

b. **Emergence of Scientific Research:** The 20th century witnessed the rise of scientific research in watershed management, driven by advances in hydrology, soil science, ecology, and related disciplines.

   Scientific studies on watershed hydrology, soil erosion processes, vegetation dynamics, and ecosystem services provided a deeper understanding of watershed dynamics and influenced management practices.

c. **Technological Advancements:** Technological innovations, such as remote sensing, geographic information systems (GIS), and computer modeling, revolutionized watershed management by enabling spatial analysis, monitoring, and predictive modeling of watershed processes.

   Remote sensing technologies provide valuable data on land cover changes, vegetation health, and hydrological patterns, facilitating informed decision-making and resource allocation in watershed management.

d. **Interdisciplinary Collaboration:** Modern watershed management approaches emphasize interdisciplinary collaboration among scientists, engineers, policymakers, and local stakeholders to address complex environmental challenges.

   Integrated watershed modeling frameworks, such as Soil and Water Assessment Tool (SWAT) and Hydrologic Engineering Center's Hydrologic Modeling System (HEC-HMS), integrate hydrological, ecological, and socio-economic components to support decision-making in watershed management.

e. **Adoption of Best Management Practices:** Evidence-based best management practices (BMPs) derived from scientific research and field trials are widely adopted in modern watershed management to address specific issues such as soil erosion, water pollution, and habitat degradation.

   BMPs include conservation tillage, riparian buffer zones, streambank stabilization, reforestation, and wetland restoration, which are implemented based on site-specific conditions and environmental objectives.

f. **Policy Integration and Adaptive Management:** Modern watershed management approaches advocate for policy integration, adaptive management, and stakeholder engagement to address emerging challenges such as climate change, land-use change, and water scarcity.

   Adaptive management frameworks emphasize continuous monitoring, evaluation, and adjustment of management strategies based on feedback from stakeholders and changing environmental conditions.

Government Policies and Initiatives Shaping Integrated Watershed Management

Government policies and initiatives play a crucial role in promoting integrated watershed management (IWM) by providing the necessary legal frameworks, institutional mechanisms, and financial support to facilitate conservation and restoration efforts. Overall, government policies and initiatives play a critical role in shaping integrated watershed management by providing policy support, institutional frameworks, and financial resources to enable sustainable natural resource management, rural development, and climate resilience in watersheds across India. This subsection explores the key policies and initiatives that have shaped IWM:

a. **National Watershed Development Program (NWDP):** NWDP, launched in 1986 by the Government of India, aimed to address soil erosion, land degradation, and water scarcity through watershed-based interventions.

The program focused on community participation, soil and water conservation measures, afforestation, and livelihood enhancement activities to improve the productivity and sustainability of rainfed agriculture in watersheds across the country.

b. **Integrated Watershed Management Program (IWMP):** IWMP, launched in 2009 as a successor to NWDP, aimed to promote holistic and participatory watershed management approaches.

The program emphasized convergence between various government departments, decentralized planning, community mobilization, and sustainable land and water management practices to enhance livelihoods and resilience in rural areas.

c. **National Rural Employment Guarantee Act (NREGA):** NREGA, enacted in 2005, provides a legal guarantee for 100 days of wage employment per year to rural households in India.

NREGA has been instrumental in supporting watershed management activities by providing employment opportunities for labor-intensive activities such as soil conservation, water harvesting, and afforestation, thereby contributing to rural development and natural resource conservation.

Figure 4. National Rural Employment Guarantee Act (NREGA)

d. **Pradhan Mantri Krishi Sinchayee Yojana (PMKSY):** PMKSY, launched in 2015, aims to improve water use efficiency in agriculture through various components, including watershed development, micro-irrigation, and water resource conservation.

The watershed component of PMKSY focuses on promoting integrated watershed management practices, rainwater harvesting, groundwater recharge, and soil moisture conservation to enhance agricultural productivity and water security.

e. **National Mission for Sustainable Agriculture (NMSA):** NMSA, launched in 2014, aims to promote sustainable agriculture practices, enhance resilience to climate change, and improve soil health and water management.

   The mission supports watershed management activities such as soil conservation, water harvesting, agroforestry, and integrated farming systems to address environmental degradation and promote climate-smart agriculture.

f. **State-Level Initiatives and Programs:** Many states in India have also implemented their own watershed development programs and initiatives, tailored to local agro-climatic conditions, resource availability, and socio-economic priorities.

   State-level initiatives complement national programs by providing additional resources, technical support, and institutional capacity building for watershed management at the grassroots level.

1.4 Principles and Objectives of Integrated Watershed Management

- Core Principles Guiding Integrated Watershed Management Integrated watershed management (IWM) is guided by a set of fundamental principles that underpin its approach to sustainable natural resource management, community empowerment, and ecosystem resilience. By adhering to these core principles, integrated watershed management endeavors to achieve sustainable development, environmental conservation, and socio-economic well-being within watersheds, fostering resilience and prosperity for present and future generations. This section outlines the core principles that inform decision-making processes and management actions in IWM:

  Sustainability: Sustainability is a central principle of IWM, emphasizing the need to balance ecological, social, and economic considerations to meet the needs of current and future generations.

  IWM aims to promote sustainable land use practices, conserve natural resources, and maintain ecosystem integrity within watersheds while ensuring the long-term viability of livelihoods and ecosystem services.

Equity and Social Justice

- Equity and social justice are essential principles in IWM, recognizing the rights and interests of all stakeholders, particularly marginalized and vulnerable communities.
- IWM endeavors to promote equitable access to water and land resources, empower local communities, and address socio-economic disparities through inclusive decision-making processes and benefit-sharing mechanisms.

Participation and Stakeholder Engagement

Participation and stakeholder engagement are core principles of IWM, emphasizing the importance of involving local communities, indigenous peoples, civil society organizations, and other stakeholders in planning, implementation, and monitoring activities.

IWM fosters participatory decision-making, knowledge sharing, and capacity building to empower stakeholders, enhance ownership, and promote collective action for sustainable watershed management.

Integrated and Interdisciplinary Approach

IWM adopts an integrated and interdisciplinary approach, recognizing the complex interactions between soil, water, vegetation, wildlife, and human activities within watersheds.

IWM integrates scientific knowledge, traditional wisdom, and local expertise to develop holistic management strategies that address multiple environmental, social, and economic dimensions of watershed management.

Adaptive Management

Adaptive management is a key principle of IWM, acknowledging the dynamic and uncertain nature of ecosystems and socio-economic systems.

IWM embraces flexibility, learning, and continuous improvement by monitoring environmental changes, evaluating management outcomes, and adjusting strategies based on feedback and new information to enhance resilience and effectiveness.

Ecosystem-based Approach

IWM adopts an ecosystem-based approach, recognizing the interconnectedness and interdependence of ecological processes, functions, and services within watersheds.

IWM aims to maintain and enhance ecosystem health, resilience, and functionality by conserving biodiversity, restoring habitats, and promoting sustainable land and water management practices that support ecosystem services.

- **Objectives Aimed at Sustainable Development and Environmental Sustainability:** Integrated watershed management (IWM) aims to achieve sustainable development and environmental sustainability by addressing various environmental, social, and economic challenges within watersheds. By pursuing these objectives, integrated watershed management endeavors to achieve a harmonious balance between environmental conservation, socio-economic development, and community well-being, thereby contributing to the sustainable development goals of both present and future generations within watersheds. The following objectives guide efforts towards these overarching goals:

a. **Conservation of Natural Resources:** Protecting and conserving soil, water, vegetation, and wildlife habitats within watersheds to maintain ecosystem integrity and biodiversity.

   Implementing soil conservation measures, water harvesting techniques, afforestation programs, and habitat restoration projects to mitigate environmental degradation and promote ecological resilience.

b. **Enhancement of Water Resources:** Ensuring sustainable water management practices that optimize water availability, quality, and distribution for human needs, agricultural productivity, and ecosystem functioning.

   Promoting water conservation, rainwater harvesting, groundwater recharge, and efficient irrigation techniques to address water scarcity, reduce water pollution, and enhance water security within watersheds.

c. **Improvement of Livelihoods and Socio-Economic Well-being:** Enhancing livelihood opportunities, income generation, and food security for local communities dependent on watershed resources through sustainable agricultural practices, alternative livelihoods, and value-added enterprises.

   Empowering marginalized and vulnerable groups, including women, indigenous peoples, and smallholder farmers, by promoting equitable access to resources, enhancing social capital, and strengthening local institutions.

d. **Climate Resilience and Disaster Risk Reduction:** Building resilience to climate change impacts, including droughts, floods, and extreme weather events, by implementing climate-adaptive practices, such as agroforestry, soil moisture conservation, and early warning systems.

Integrating climate resilience and disaster risk reduction measures into watershed management planning, infrastructure development, and community-based adaptation strategies to minimize vulnerability and enhance adaptive capacity.

e. **Promotion of Sustainable Agriculture and Land Use:** Adopting sustainable agricultural practices, including conservation agriculture, organic farming, and agroecological approaches, to improve soil health, enhance productivity, and reduce environmental impacts such as soil erosion and chemical runoff.

Promoting land-use planning, zoning regulations, and land tenure reforms that balance competing land uses, protect critical habitats, and minimize land degradation within watersheds.

f. **Community Empowerment and Participation:** Fostering community empowerment, social inclusion, and participatory decision-making processes that engage local stakeholders in planning, implementing, and monitoring watershed management activities.

Strengthening local institutions, building capacity, and promoting knowledge exchange to enhance local ownership, governance effectiveness, and sustainable natural resource management outcomes.

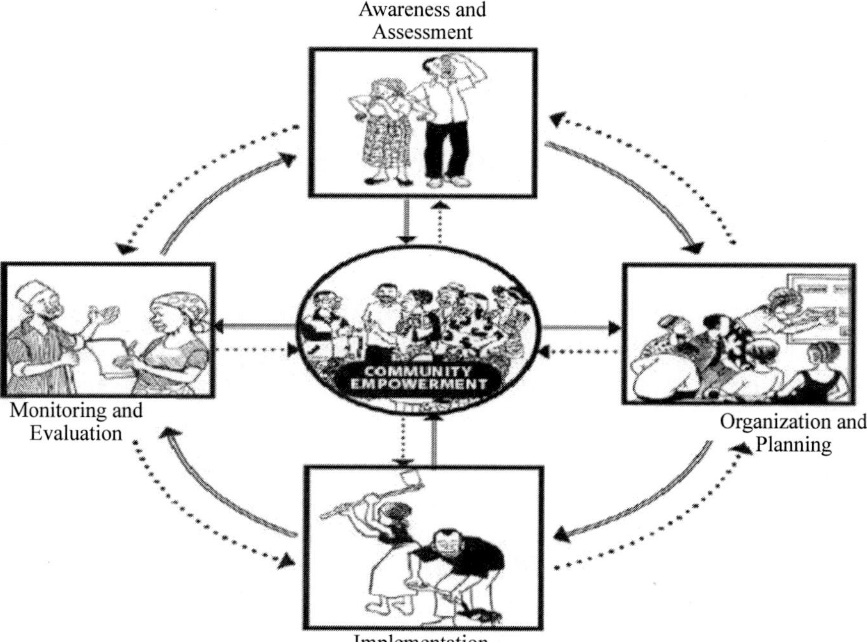

Figure 5. Community empowerment and participation through learning

Role of Stakeholder Collaboration and Community Participation

Stakeholder collaboration and community participation are essential pillars of integrated watershed management (IWM), facilitating collective action, shared decision-making, and sustainable outcomes. By fostering stakeholder collaboration and community participation, integrated watershed management can harness the collective wisdom, resources, and actions of diverse actors to achieve shared goals of environmental sustainability, socio-economic development, and resilience within watersheds. This section elucidates the significance of stakeholder engagement and community participation in achieving the objectives of IWM:

a. **Inclusive Decision-Making Processes:** Stakeholder collaboration ensures that diverse perspectives, interests, and priorities are considered in the planning, implementation, and evaluation of watershed management initiatives.

   Inclusive decision-making processes empower local communities, indigenous peoples, women, and marginalized groups to contribute their knowledge, values, and needs, fostering ownership and legitimacy of management actions.

b. **Enhanced Resource Mobilization and Allocation:** Collaborative partnerships among government agencies, non-governmental organizations (NGOs), research institutions, and private sector entities facilitate resource mobilization and allocation for watershed management activities.

   Stakeholder collaboration enables pooling of financial resources, technical expertise, and institutional capacities to address complex environmental challenges and scale-up conservation efforts within watersheds.

c. **Holistic and Integrated Approaches:** Stakeholder collaboration promotes holistic and integrated approaches to watershed management by fostering cooperation across sectors, disciplines, and administrative boundaries.

   Collaborative efforts leverage synergies between water, agriculture, forestry, energy, and biodiversity sectors to address interconnected challenges such as water scarcity, soil erosion, biodiversity loss, and climate change impacts within watersheds.

d. **Capacity Building and Knowledge Sharing:** Community participation in watershed management initiatives enhances local capacity, knowledge, and skills for sustainable natural resource management.

Collaborative learning platforms, capacity-building workshops, and knowledge exchange networks facilitate the sharing of best practices, indigenous wisdom, and scientific insights among stakeholders, empowering communities to make informed decisions and adopt innovative solutions.

e. **Conflict Resolution and Consensus Building:** Stakeholder collaboration fosters dialogue, negotiation, and consensus building to resolve conflicts, address competing interests, and reconcile divergent viewpoints within watersheds.

   Participatory decision-making processes build trust, social cohesion, and mutual understanding among stakeholders, leading to more equitable and durable solutions to watershed management challenges.

f. **Community Empowerment and Ownership:** Community participation in watershed management initiatives promotes empowerment, self-reliance, and local ownership of conservation efforts.

   Engaging local communities in planning, implementing, and monitoring activities fosters a sense of responsibility, pride, and stewardship towards watershed resources, leading to sustained behavioral change and long-term sustainability.

Important Questions

1. Why are institutional frameworks and governance mechanisms essential for effective watershed management?
2. Can you explain the role of government agencies in watershed management, and provide examples of specific agencies involved in this field?
3. How do community-based organizations contribute to watershed management, and what are some examples of their involvement in specific projects or initiatives?
4. What are stakeholder engagement mechanisms, and why are they important in watershed management decision-making processes?
5. How can partnerships and collaborations enhance the effectiveness of watershed management initiatives? Can you provide examples of successful partnerships in this context?
6. What are some key components of legal and policy frameworks that support watershed management, and how do they influence decision-making and implementation?

7. How do monitoring and evaluation mechanisms contribute to the success of watershed management interventions, and what are some common indicators used to assess progress?
8. Why is capacity building and training important for stakeholders involved in watershed management, and what are some examples of capacity-building initiatives in this field?
9. What are some of the challenges that integrated watershed management in India aims to address?
10. How has the concept of integrated watershed management evolved over time in India?
11. What are some of the adverse impacts of human activities on watersheds mentioned in the text?
12. How does watershed management contribute to sustainable water resource management?
13. What role do technological advancements play in fostering resilient watershed ecosystems?
14. How does the book aim to explore the complexities of integrated watershed management in India?
15. Can you elaborate on the concept of a watershed and its significance in environmental conservation?
16. Why is it important to adopt a holistic approach in watershed management?
17. What are some potential future prospects for integrated watershed management in India?
18. How does integrated watershed management incorporate climate change considerations into its planning and implementation processes?
19. What are some key aspects of climate-resilient agriculture mentioned in the text, and how do they contribute to reducing vulnerability to climate variability and extremes?
20. How does strengthening water resource management systems contribute to coping with changing precipitation patterns and increased variability in water availability?
21. In what ways does empowering local communities contribute to effective climate adaptation measures in watershed management?
22. What are the benefits of ecosystem restoration and conservation in enhancing resilience to climate change impacts, as mentioned in the text?

23. How do early warning systems and disaster preparedness mechanisms help in reducing the loss of lives and livelihoods from climate-related hazards?
24. How can policy frameworks and institutional support help in mainstreaming climate resilience and risk reduction considerations into watershed management plans and programs?
25. What role does research, innovation, and knowledge-sharing platforms play in promoting climate adaptation and risk reduction in watershed management?

2

Characteristics of Watershed

2.1 Definition and Boundaries of Watershed Areas

Concept of Watershed Boundaries

The concept of watershed boundaries is fundamental to understanding the hydrological and ecological dynamics of a given landscape. Watershed boundaries, also known as drainage divides or catchment areas, delineate the geographical extent of a watershed, which is an area of land where all the water drains to a common outlet, such as a stream, river, lake, or ocean. Watershed boundaries represent the natural hydrological and ecological units of a landscape, reflecting the interconnectedness of water resources and ecosystems. Understanding the factors influencing their definition is essential for effective watershed management, resource planning, and environmental conservation. Exploring the delineation of watershed boundaries involves considering various factors that influence their definition:

a. **Topography:** Topography, particularly the elevation and slope of the land, plays a significant role in defining watershed boundaries. Watershed boundaries typically follow ridgelines or high points in the landscape, where water flows in different directions, ultimately converging into distinct drainage networks.

b. **Hydrology:** Hydrological characteristics, such as precipitation patterns, surface runoff, and groundwater flow, influence the delineation of watershed boundaries. Areas with higher rainfall or snowmelt may have larger watersheds with more extensive drainage areas, while areas with low precipitation may have smaller, more localized watersheds.

c. **Geomorphology:** Geomorphological features, such as valleys, hills, and geological formations, can affect the shape and extent of watershed boundaries. Geological structures and soil types may influence the movement and storage of water within a watershed, influencing the delineation of drainage basins.

d. **Land Use and Land Cover:** Human activities and land use practices, such as agriculture, urbanization, and deforestation, can alter the natural flow of water and affect watershed boundaries. Changes in land cover can impact the infiltration rate, runoff patterns, and water quality within a watershed, necessitating adjustments to watershed delineation.

e. **Vegetation and Ecology**: Vegetation types, biodiversity, and ecological processes also influence watershed boundaries. Forested areas, wetlands, and riparian zones play important roles in regulating water flow, reducing erosion, and maintaining water quality, which can influence the delineation of watershed boundaries.

f. **Anthropogenic Factors:** Human interventions, such as construction of dams, reservoirs, and drainage systems, can modify natural drainage patterns and alter watershed boundaries. Anthropogenic activities may create artificial divides or alter the flow paths of water, requiring adjustments to watershed delineation.

g. **Scale and Resolution:** The scale and resolution of data used for watershed delineation can impact the accuracy and precision of boundary delineation. Higher-resolution topographic data, satellite imagery, and hydrological models can provide more detailed information for delineating watershed boundaries at different scales.

Methods of Watershed Delineation

Certainly! Various techniques are used to delineate watershed boundaries, each offering unique advantages and capabilities. Each of these methods has its strengths and limitations, and the choice of technique depends on factors such as data availability, project objectives, spatial scale, and computational resources. Combining multiple methods and validation techniques can enhance the accuracy and reliability of watershed delineation efforts, supporting informed decision-making in watershed management and environmental planning. Here's a discussion of some common methods:

a. **Manual Delineation:** Historically, watershed boundaries were delineated manually using topographic maps, aerial photographs, and field surveys. Cartographers would trace the ridgelines or high points on topographic maps to define the boundaries of watersheds. While this method is labor-intensive and time-consuming, it provides detailed control over the delineation process and can be useful in areas with limited data availability.

b. **Topographic Maps:** Traditional topographic maps, which depict elevation contours, landforms, and hydrological features, remain valuable tools for watershed delineation. By identifying the highest points or ridgelines in the landscape, cartographers can trace the drainage divides to delineate watershed boundaries. While topographic maps provide visual insights into the terrain, their accuracy and resolution may vary depending on the scale of the map.

c. **Digital Elevation Models (DEMs):** Digital Elevation Models, derived from satellite imagery or aerial LiDAR (Light Detection and Ranging) data, provide detailed representations of terrain elevation in raster format. DEMs are widely used for watershed delineation due to their high resolution and accuracy. Common algorithms, such as the D8 algorithm, use DEMs to calculate flow directions and delineate watersheds based on elevation gradients and drainage patterns.

d. **Geographic Information Systems (GIS):** GIS software enables the integration of various spatial datasets and analytical tools for watershed delineation. GIS-based approaches utilize DEMs, hydrological models, and spatial analysis techniques to automate the process of watershed delineation. GIS allows for the efficient processing of large datasets, customization of analysis parameters, and visualization of watershed boundaries in digital maps.

e. **Hydrological Modelling:** Hydrological models, such as the Soil and Water Assessment Tool (SWAT) or the Hydrologic Engineering Center's Hydrologic Modeling System (HEC-HMS), simulate the movement of water across the landscape based on rainfall, soil properties, land use, and topography. These models can be used to predict streamflow, identify drainage networks, and delineate watersheds by defining contributing areas to specific outlets or pour points.

f. **Automated Tools and Algorithms**: Several automated tools and algorithms have been developed specifically for watershed delineation, including the Watershed Boundary Dataset (WBD) in the United States and the HydroSHEDS dataset globally. These datasets use advanced algorithms and spatial analysis techniques to delineate watershed boundaries at various scales, providing standardized data for watershed management and research purposes.

Importance of Watershed Boundaries

The importance of clearly defined watershed boundaries cannot be overstated when it comes to effective watershed management and planning. It is clearly defined watershed boundaries are essential for effective watershed management and planning, providing a spatial context for understanding hydrological processes, managing natural resources, protecting water quality, reducing disaster risks, conserving biodiversity, and fostering community engagement.

Recognizing the significance of watershed boundaries enables stakeholders to adopt holistic and integrated approaches that promote sustainable development and resilience in the face of environmental change. Here's a closer examination of their significance:

a. **Hydrological Understanding:** Defined watershed boundaries delineate the geographical extent of an area where all surface water drains to a common outlet, such as a stream, river, lake, or ocean. Understanding watershed boundaries is essential for comprehending the hydrological dynamics of a landscape, including surface runoff patterns, groundwater recharge areas, and flow pathways. This knowledge forms the basis for water resources management, flood control, drought mitigation, and water quality protection initiatives.

b. **Natural Resource Management:** Watershed boundaries serve as natural units for managing and conserving land, water, and ecosystems. By delineating the boundaries of a watershed, resource managers can identify and prioritize areas for conservation, restoration, and sustainable land use practices. Watershed-based management approaches promote integrated strategies that consider the interconnectedness of ecological processes and human activities within a defined geographical area.

c. **Pollution Control and Water Quality Management**: Clearly defined watershed boundaries facilitate the identification of pollution sources and the implementation of targeted measures to improve water quality. By understanding the upstream sources of contaminants and their transport pathways within a watershed, managers can develop strategies for pollution prevention, erosion control, riparian buffer establishment, and wetland restoration to safeguard water resources and aquatic ecosystems.

d. **Floodplain Management and Disaster Risk Reduction**: Watershed boundaries delineate the extent of floodplains and areas prone to flooding during extreme weather events. Effective floodplain management relies on accurate delineation of watershed boundaries to assess flood risks, develop floodplain zoning regulations, and implement flood mitigation measures such as levees, floodwalls, and natural floodplain restoration projects. Clear identification of watershed boundaries enhances community resilience to floods and reduces vulnerability to disaster risks.

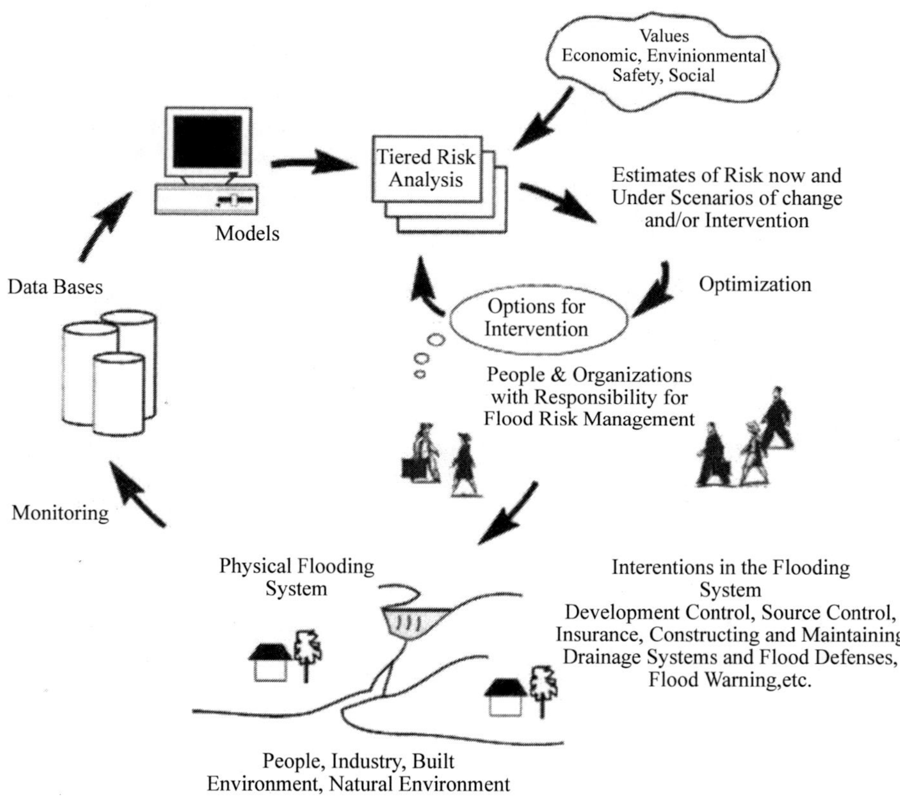

Figure 6. Flood risk management cycle by Paul Sayers (2023)

e. **Ecosystem Services and Biodiversity Conservation**: Watershed boundaries encompass diverse ecosystems and habitats that provide essential ecosystem services, such as water purification, habitat provision, carbon sequestration, and recreational opportunities. Protecting intact watersheds and conserving biodiversity within watershed boundaries is crucial for maintaining ecosystem health, supporting wildlife habitats, and sustaining human well-being.

f. **Integrated Planning and Decision-Making:** Watershed-based planning facilitates integrated approaches to land use planning, infrastructure development, and natural resource management across administrative boundaries. By considering the interconnectedness of social, economic, and environmental factors within a watershed, planners can develop holistic strategies that balance competing interests, minimize conflicts, and promote sustainable development while preserving ecosystem integrity.

g. **Community Engagement and Stakeholder Collaboration:** Watershed boundaries provide a spatial framework for engaging local communities, stakeholders, and decision-makers in participatory watershed management processes. By fostering collaboration and collective action at the watershed scale, stakeholders can work together to address shared challenges, build consensus, and implement solutions that benefit both people and nature.

2.2 Topography and Geomorphology of Watersheds

Characteristics of Watershed Topography

The physical characteristics of watershed topography play a crucial role in shaping hydrological processes and influencing the behavior of water within a watershed. The characteristics of watershed topography is essential for assessing hydrological processes, predicting water movement, and managing water resources effectively. By considering elevation, slope, drainage patterns, channel networks, watershed size and shape, and land use characteristics, watershed managers can develop strategies to mitigate flood risk, conserve water resources, protect ecosystems, and promote sustainable land use practices within a watershed. Here's a description of some key characteristics and their influence:

a. **Elevation:** Elevation refers to the height above sea level and varies across the landscape within a watershed. High-elevation areas, such as mountains or ridgelines, often serve as watershed boundaries, where water flows downhill in different directions to form distinct drainage basins. Elevation gradients influence the velocity and direction of water flow, as water moves from higher to lower elevations, driving hydrological processes such as runoff, infiltration, and groundwater recharge.

b. **Slope:** Slope represents the steepness or inclination of the land surface and affects the rate of water movement and erosion within a watershed. Steeper slopes promote faster runoff and increased erosion, as gravity pulls water downhill more quickly, leading to higher stream velocities and sediment transport. Gentle slopes, on the other hand, allow for slower runoff and greater infiltration, facilitating groundwater recharge and reducing erosion potential.

c. **Drainage Patterns:** Drainage patterns, determined by the arrangement of streams and rivers within a watershed, reflect the underlying geology, topography, and land use characteristics. Common drainage patterns include dendritic (tree-like), parallel, trellis, radial, and rectangular patterns, each associated with different geological settings and landscape features. Drainage patterns influence the flow pathways and connectivity of water within a watershed, affecting flood risk, sediment transport, and water quality.

d. **Channel Networks:** Channel networks within watersheds consist of streams, rivers, and tributaries that convey surface runoff and provide pathways for water flow. The hierarchical arrangement of channels, ranging from small headwater streams to larger rivers, influences the distribution of water and sediment throughout the watershed. Channel morphology, including channel width, depth, and sinuosity, affects flow dynamics, habitat suitability, and erosion rates within stream channels.

e. **Watershed Size and Shape:** The size and shape of a watershed, often expressed as its drainage area or catchment size, determine the total volume of water that contributes to the watershed and the length of flow paths from upstream to downstream areas. Larger watersheds tend to have longer flow paths and higher flow volumes, while smaller watersheds may respond more rapidly to rainfall events. Watershed shape also influences the timing and magnitude of peak flows, with elongated watersheds potentially experiencing more prolonged runoff and flood events.

f. **Land Use and Land Cover:** The distribution of land use and land cover types within a watershed, including forests, agriculture, urban areas, wetlands, and impervious surfaces, affects hydrological processes such as infiltration, evapotranspiration, and surface runoff. Vegetation cover influences water retention, erosion control, and nutrient cycling, while impervious surfaces increase runoff and reduce infiltration, leading to higher flood risk and degraded water quality.

Geomorphological Processes

Geomorphological processes play a significant role in shaping watershed landscapes, influencing landforms, soil characteristics, and hydrological dynamics. The interplay of geological and geomorphological processes shapes the physical characteristics of watershed landscapes, influencing their hydrological behavior, erosion susceptibility, sediment transport dynamics, and ecological functioning. Understanding these processes is essential for watershed management, land use planning, and natural hazard mitigation efforts aimed at promoting resilience and sustainability in watershed environments. Here's an exploration of the geological and geomorphological factors that contribute to watershed evolution:

a. **Erosion:** Erosion is the process of wearing away rock and soil by the action of water, wind, ice, or gravity. In watersheds, water erosion is a dominant force, particularly during rainfall events and snowmelt. Running water erodes soil and bedrock through processes such as splash erosion, sheet erosion, rill erosion, and gully erosion. Erosion rates vary depending on factors such as slope steepness, soil type, vegetation cover, and land use practices. Erosion removes sediment from hillslopes and transports it downstream, shaping the landscape over time.

b. **Sedimentation:** Sedimentation is the deposition of eroded material, including sand, silt, and clay, in river channels, floodplains, and deltaic environments. Sedimentation occurs when the energy of flowing water decreases, allowing sediment particles to settle out of suspension and accumulate on the landscape. In watersheds, sedimentation processes contribute to the formation of alluvial deposits, riverbanks, floodplains, and deltaic landforms. Excessive sedimentation can impair water quality, degrade aquatic habitats, and increase flood risk in downstream areas.

c. **Fluvial Processes:** Fluvial processes, related to the flow of water in rivers and streams, shape watershed landscapes through erosion, transportation, and deposition of sediment. River channels exhibit dynamic behavior, adjusting their size, shape, and gradient in response to changes in discharge, sediment load, and base level. Fluvial processes create landforms such as meanders, braided channels, oxbow lakes, river terraces, and alluvial fans, which reflect the interaction between water, sediment, and underlying geology.

d. **Glacial Processes:** Glacial processes, associated with the movement of glaciers and ice sheets, have sculpted many watershed landscapes, particularly in regions affected by past glaciation. Glaciers erode bedrock through processes such as plucking and abrasion, creating U-shaped valleys, cirques, moraines, and glacial lakes. As glaciers retreat, they leave behind deposits of till, gravel, and sand, shaping the topography and hydrology of glaciated watersheds.

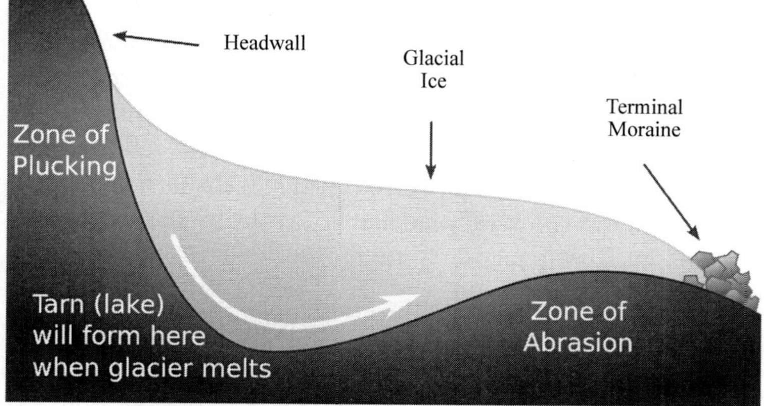

Figure 7. Different glacial processes

e. **Weathering and Mass Movement:** Weathering processes, including mechanical weathering, chemical weathering, and biological weathering, break down rock and minerals into smaller particles, contributing to soil formation and landscape evolution. Mass movement processes, such as landslides, rockfalls, and soil creep, redistribute material downslope under the influence of gravity, altering the topography and stability of hillslopes within watersheds.

f. **Tectonic Processes:** Tectonic processes, associated with the movement of Earth's lithospheric plates, influence watershed landscapes through crustal deformation, faulting, and uplift. Tectonic activity can create mountain ranges, rift valleys, and structural basins, shaping the overall relief and drainage patterns of watersheds. Faults and fractures in the Earth's crust may act as conduits for groundwater flow and influence the distribution of mineral resources within watersheds.

Landform Classification

Landforms within watersheds exhibit diverse shapes, characteristics, and functions, reflecting the complex interactions of geological, geomorphological, and hydrological processes. The characteristics and functions of different landforms within watersheds is essential for effective watershed management, land use planning, and natural resource conservation efforts. By recognizing the role of landforms in regulating water flow, sediment transport, and ecosystem functioning, watershed managers can develop strategies to enhance watershed resilience, protect valuable habitats, and promote sustainable use of natural resources for present and future generations. Here's an introduction to different types of landforms found within watersheds and their implications for watershed management:

a. **Valleys:** Valleys are low-lying areas between hills or mountains, often formed by fluvial erosion and sedimentation processes. Valleys serve as conduits for water flow, collecting runoff from surrounding uplands and channelling it downstream. Valleys may include features such as floodplains, terraces, meanders, and alluvial fans, which influence the distribution of water, sediment, and habitats within watersheds. Effective management of valleys involves balancing floodplain protection, habitat restoration, and land use planning to minimize flood risk and preserve ecological integrity.

b. **Hills and Ridges**: Hills and ridges are elevated landforms that define the topography of watersheds, separating drainage basins and influencing water flow patterns. Hills and ridges may be composed of bedrock, soil, or glacial deposits, exhibiting varying degrees of slope steepness and vegetation cover. These landforms contribute to runoff generation, groundwater recharge, and erosion susceptibility within watersheds. Sustainable management of hills and ridges involves practices such as soil conservation, reforestation, and slope stabilization to reduce erosion, enhance water retention, and protect biodiversity.

c. **Mountains:** Mountains are prominent landforms characterized by steep slopes, rugged terrain, and high elevation, often found at the headwaters of watersheds. Mountains play a critical role in regulating water supply, as they capture precipitation, store snowpack, and release water gradually over time. Mountainous watersheds provide freshwater resources for downstream communities, support unique ecosystems, and offer recreational opportunities. Effective management of mountainous watersheds involves balancing water allocation, ecosystem conservation, and sustainable tourism development to preserve mountain ecosystems and ensure water security for downstream users.

d. **Basins and Depressions**: Basins and depressions are concave landforms that collect water and sediment within watersheds, forming localized drainage areas. Basins may include features such as lakes, wetlands, ponds, and marshes, which provide important ecosystem services such as water storage, flood attenuation, and habitat provision. Basins and depressions play a crucial role in regulating water quality, nutrient cycling, and biodiversity within watersheds. Management strategies for basins and depressions focus on enhancing their resilience to climate change, protecting critical habitats, and promoting sustainable land use practices to maintain ecosystem functions and services.

e. **Plateaus and Mesas:** Plateaus and mesas are elevated flat-topped landforms with steep cliffs or escarpments, often found in arid or semi-arid regions. These landforms may contain valuable groundwater resources, fossil deposits, and mineral resources, making them significant for both ecological and economic reasons. Plateaus and mesas influence water infiltration, recharge rates, and groundwater flow paths within watersheds. Sustainable management of plateaus and mesas involves conservation of unique ecosystems, sustainable groundwater extraction, and responsible mineral resource development to balance environmental and economic priorities.

2.3 Hydrological Cycle and Water Flow Patterns

Hydrological Processes

Hydrological processes are fundamental to the movement, distribution, and storage of water within watersheds, influencing water availability, quality, and ecosystem functioning. These hydrological processes interact dynamically within watersheds, forming a complex cycle that regulates the movement and distribution of water across terrestrial and aquatic ecosystems. Understanding the dynamics of the hydrological cycle is essential for watershed management, water resources planning, flood control, drought mitigation, and ecosystem conservation efforts aimed at sustaining water security, environmental resilience, and human well-being within watersheds. Here's an overview of the key components of the hydrological cycle within watersheds:

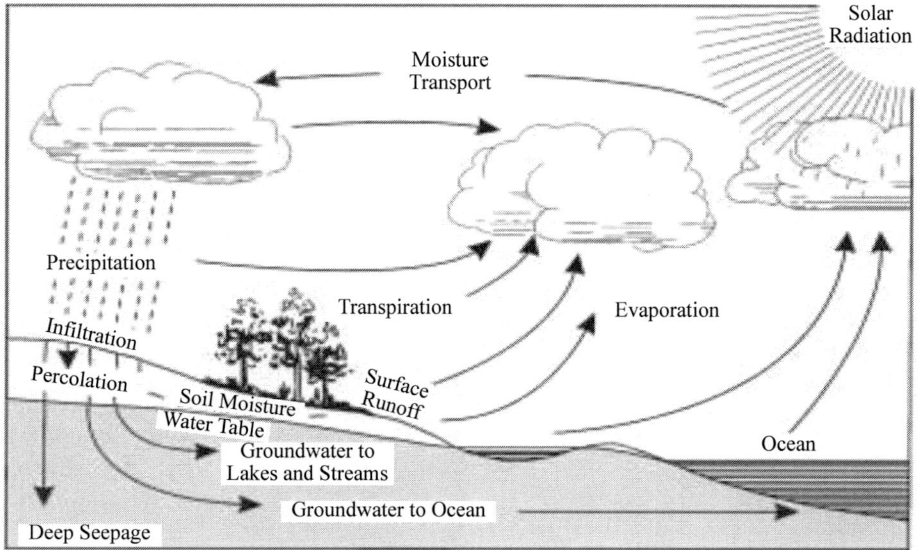

Figure 8. Hydrological cycle

a. **Precipitation:** Precipitation is the primary input of water into watersheds, encompassing rainfall, snowfall, sleet, and hail. Precipitation occurs when moisture-laden air masses cool and condense, releasing moisture in the form of droplets or ice crystals. The distribution, intensity, and frequency of precipitation events vary spatially and temporally within watersheds, influencing water availability for downstream users, vegetation, and ecosystems.

b. **Runoff:** Runoff is the portion of precipitation that flows over the land surface, eventually reaching streams, rivers, lakes, or oceans. Runoff occurs when precipitation exceeds the infiltration capacity of soil, or when the soil becomes saturated due to prolonged rainfall or snowmelt. Runoff pathways may include surface flow over impermeable surfaces, subsurface flow through soil layers, and channelized flow in streams and rivers. Runoff plays a critical role in replenishing surface water bodies, sustaining aquatic habitats, and supporting human activities such as agriculture, industry, and recreation.

c. **Infiltration:** Infiltration is the process by which precipitation penetrates the soil surface and percolates into the subsurface layers, replenishing soil moisture and groundwater reserves. Infiltration rates depend on soil properties, such as texture, structure, porosity, and organic matter content, as well as land use practices and vegetation cover. Infiltrated water may contribute to groundwater recharge, subsurface flow, and baseflow in streams, enhancing water storage and resilience within watersheds.

d. **Evaporation:** Evaporation is the transfer of water from the Earth's surface to the atmosphere in the form of vapor, driven by solar radiation, temperature, humidity, and wind speed. Evaporation occurs from various surfaces within watersheds, including soil, water bodies, vegetation, and impervious surfaces. Evapotranspiration, which combines evaporation from soil and water surfaces with transpiration from plants, represents a significant loss of water from watersheds, affecting water balance, ecosystem productivity, and climate regulation.

e. **Transpiration:** Transpiration is the process by which plants absorb water from the soil through their roots and release it into the atmosphere through stomata in their leaves. Transpiration rates vary depending on plant species, leaf characteristics, soil moisture availability, and environmental conditions such as temperature and humidity. Transpiration plays a crucial role in regulating plant water status, nutrient uptake, and photosynthesis, as well as influencing local and regional climate patterns through the release of water vapor into the atmosphere.

Surface Water Flow Patterns

Surface water flow patterns within watersheds are influenced by various factors, including topography, geology, land cover, precipitation, and human activities. Surface water flow patterns within watersheds are dynamic and complex, reflecting the interplay of natural processes and human activities across diverse landscapes. Understanding these flow patterns is essential for watershed management, water resources planning, flood control, habitat restoration, and ecosystem conservation efforts aimed at sustaining the health and resilience of aquatic ecosystems and human communities within watersheds. Here's an examination of the pathways and dynamics of surface water flow within watersheds:

a. **Topography:** Topography plays a fundamental role in shaping surface water flow patterns within watersheds. Water flows downhill under the influence of gravity, following the natural slope of the land surface. Hills, ridges, valleys, and other topographic features define the direction and intensity of surface water flow, guiding the formation of drainage networks and channel networks within watersheds.

b. **Drainage Basins:** Drainage basins, also known as catchments or watersheds, are the primary units of surface water flow within a landscape. A drainage basin is an area of land where all surface water drains to a common outlet, such as a stream, river, lake, or ocean. Drainage basins are delineated by topographic divides, such as ridgelines or high points, which separate one drainage basin from another. The boundaries of drainage basins define the extent of surface water flow and influence the distribution of water resources within watersheds.

c. **Stream Channels:** Stream channels are the pathways through which surface water flows within watersheds, forming interconnected networks of streams, rivers, and tributaries. Stream channels vary in size, shape, and morphology, ranging from small headwater streams to large rivers. Stream channels may exhibit features such as meanders, riffles, pools, and rapids, which reflect the interaction between water flow, sediment transport, and channel morphology. Stream channels play a critical role in conveying surface runoff, regulating water quality, and providing habitat for aquatic organisms within watersheds.

d. **Channel Morphology:** Channel morphology refers to the physical characteristics of stream channels, including channel width, depth, slope, sinuosity, and bed material composition. Channel morphology is influenced by factors such as streamflow regime, sediment supply, bank stability, and human disturbances. Natural processes such as erosion, sedimentation, and meandering shape channel morphology over time, creating diverse habitats and geomorphic features within watersheds.

e. **Hydrological Connectivity:** Hydrological connectivity refers to the degree to which surface water flows are interconnected within a watershed, facilitating the exchange of water, sediment, and nutrients between different parts of the landscape. Hydrological connectivity is influenced by factors such as channel geometry, streamflow dynamics, floodplain connectivity, and land use patterns. High levels of hydrological connectivity promote water retention, biodiversity, and ecosystem resilience within watersheds, while fragmentation of flow pathways can disrupt ecosystem functioning and degrade water quality.

Groundwater Recharge and Discharge

Groundwater recharge and discharge are essential components of the hydrological cycle within watersheds, influencing water availability, groundwater storage, and surface water interactions. The interactions between groundwater recharge and discharge within watersheds are complex and dynamic, driven by a combination of natural processes and human activities. Understanding these mechanisms is essential for sustainable water resources management, groundwater protection, and ecosystem conservation efforts aimed at maintaining water security and resilience within watersheds. Here's an exploration of the mechanisms of groundwater recharge and discharge and their interactions with surface water bodies within watersheds:

a. **Groundwater Recharge:** Groundwater recharge refers to the process by which water infiltrates the soil surface and percolates downward to replenish underground aquifers. Several mechanisms contribute to groundwater recharge within watersheds:

b. **Infiltration:** Infiltration occurs when precipitation or surface water infiltrates the soil surface and moves vertically downward through soil pores and rock fractures, eventually reaching the water table. Infiltration rates depend on soil properties, land cover, vegetation type, and precipitation intensity.

c. **Percolation:** Percolation refers to the downward movement of water through unsaturated soil layers and into the saturated zone, where groundwater is stored in aquifers. Percolation rates are influenced by soil permeability, hydraulic conductivity, and the presence of impermeable layers or barriers.

d. **Inflow from Surface Water Bodies**: Surface water bodies such as streams, rivers, lakes, and wetlands can contribute to groundwater recharge through direct infiltration of surface water into underlying aquifers. This process, known as bank infiltration or riverbed seepage, occurs where surface water bodies intersect with groundwater flow paths.

e. **Artificial Recharge**: Artificial recharge involves intentional or engineered methods to increase groundwater replenishment, such as infiltration basins, injection wells, recharge ponds, and spreading grounds. Artificial recharge techniques can enhance groundwater storage, mitigate overdraft, and improve water quality within watersheds.

f. **Groundwater Discharge**: Groundwater discharge refers to the release of water from aquifers to surface water bodies, springs, or evapotranspiration zones. Several mechanisms contribute to groundwater discharge within watersheds:

g. **Spring Outflow**: Springs are natural outlets where groundwater emerges at the land surface, forming localized surface water features. Springs occur where groundwater intersects the ground surface along geological faults, fractures, or permeable formations, providing a source of baseflow to streams and rivers.

h. **Baseflow Contribution**: Groundwater discharge contributes to baseflow, the sustained flow of water in streams and rivers between precipitation events. Baseflow maintains streamflow during dry periods, sustains aquatic habitats, and supports ecosystem functions within watersheds.

i. **Evapotranspiration:** Groundwater contributes to evapotranspiration processes by supplying water to vegetation roots, which then release water vapor into the atmosphere through transpiration. Groundwater-supported vegetation plays a crucial role in regulating local and regional climate patterns, as well as influencing groundwater recharge rates within watersheds.

j. **Subsurface Outflow:** Subsurface outflow occurs when groundwater flows laterally through aquifers and discharges into surface water bodies or adjacent watersheds. Subsurface flow paths are influenced by geological structures, hydrological gradients, and hydraulic conductivity variations within aquifer systems.

2.4 Soil Types and Soil Erosion Dynamics

Soil Characteristics

Soil characteristics within watersheds vary widely, reflecting the diverse geological, climatic, and land use conditions across different landscapes. Understanding the soil types and properties within watersheds is essential for land use planning, agricultural management, water resources management, and ecosystem conservation efforts. By assessing soil characteristics, stakeholders can develop strategies to optimize soil fertility, water retention, and ecosystem services while minimizing environmental impacts and promoting sustainable land use practices within watersheds. Here's a description of the various soil types and properties commonly found within watersheds:

a. **Texture:** Soil texture refers to the relative proportions of sand, silt, and clay particles in the soil. Soil texture influences water retention, drainage, and fertility, as well as root penetration and aeration. Sandy soils have larger particles and higher permeability, allowing water to drain quickly but holding less moisture and nutrients. Silty soils have intermediate particle sizes and moderate water retention capacity, while clayey soils have smaller particles and higher water retention but may be prone to waterlogging and compaction.

b. **Structure:** Soil structure refers to the arrangement of soil particles into aggregates or clumps, which affects soil porosity, aeration, and root penetration. Well-structured soils have stable aggregates with pore spaces for air and water movement, promoting root growth and nutrient uptake. Soil structure can be influenced by factors such as organic matter content, microbial activity, soil moisture, and tillage practices. Poor soil structure, such as compaction or crusting, can impair water infiltration, root development, and plant productivity.

c. **Fertility:** Soil fertility refers to the ability of soil to provide essential nutrients for plant growth and development. Fertile soils contain adequate levels of macronutrients (nitrogen, phosphorus, potassium) and micronutrients (iron, zinc, copper) necessary for plant metabolism and yield. Soil fertility depends on factors such as parent material, weathering processes, organic matter content, microbial activity, and nutrient cycling. Soil fertility assessments help inform fertilizer management, crop rotation, and soil amendment practices to sustain agricultural productivity and ecosystem health within watersheds.

d. **pH:** Soil pH is a measure of the acidity or alkalinity of soil, which influences nutrient availability, microbial activity, and plant growth. Soils with pH values below 7 are acidic, while soils with pH values above 7 are alkaline. Soil pH affects the solubility of nutrients and toxic elements, with optimal pH levels promoting nutrient uptake and plant health. Soil pH can be modified through liming or acidification practices to adjust soil chemistry and improve crop performance within watersheds.

e. **Permeability:** Soil permeability refers to the ability of soil to transmit water and air through its pore spaces, influencing water infiltration, drainage, and groundwater recharge. Permeability is determined by soil texture, structure, porosity, and compaction. Sandy soils have high permeability and rapid drainage rates, while clayey soils have low permeability and slower drainage rates. Permeability assessments help identify soil management practices to enhance water infiltration, reduce runoff, and minimize erosion within watersheds.

Soil Erosion Processes

Soil erosion is a natural process influenced by various factors within watersheds, including rainfall, runoff, slope, land use practices, and vegetation cover. Understanding the mechanisms and contributors to soil erosion is essential for effective erosion control and watershed management. Soil erosion within watersheds is a complex process driven by interactions between natural and anthropogenic factors. By understanding the mechanisms and contributors to soil erosion, stakeholders can implement erosion control measures, land management practices, and conservation strategies to protect soil resources, water quality, and ecosystem integrity within watersheds. Here's an overview of the key factors contributing to soil erosion within watersheds:

a. **Rainfall Intensity and Distribution**: Rainfall is a primary driver of soil erosion, as intense or prolonged rainfall events can dislodge soil particles and generate surface runoff. High-intensity rainfall events increase the erosive power of runoff, leading to greater soil detachment and transport downslope. Rainfall distribution patterns, including frequency, duration, and intensity, influence soil erosion rates within watersheds, with areas experiencing heavy or concentrated rainfall more prone to erosion.

b. **Surface Runoff:** Surface runoff occurs when rainfall exceeds the infiltration capacity of soil, resulting in water flowing over the land surface and carrying soil particles with it. Runoff erodes soil through processes such as sheet erosion, rill erosion, and gully erosion, depending on the flow velocity and channelization of runoff pathways. Slope steepness, soil permeability, and land cover influence the generation and routing of surface runoff within watersheds, with steeper slopes and impermeable surfaces promoting faster runoff and increased erosion rates.

c. **Slope Gradient and Length:** Slope gradient and length are critical factors affecting the velocity and erosive power of runoff within watersheds. Steeper slopes accelerate surface runoff and increase soil detachment and transport downslope, leading to higher erosion rates. Longer slope lengths provide more opportunity for runoff to gain momentum and carry sediment over greater distances, exacerbating erosion processes. Slope steepness and length interact with rainfall intensity to determine the potential for soil erosion within watersheds.

d. **Land Use and Land Cover:** Land use practices and vegetation cover influence soil erosion rates within watersheds by affecting soil stability, surface roughness, and infiltration capacity. Agricultural activities such as plowing, tilling, and crop harvesting can disturb soil structure and expose bare soil surfaces, making them more susceptible to erosion. Deforestation, urbanization, and construction activities remove vegetation cover and increase impervious surfaces, reducing the protective function of vegetation and exacerbating soil erosion.

e. **Vegetation Cover and Root Systems:** Vegetation cover plays a crucial role in mitigating soil erosion by reducing the erosive impact of rainfall, stabilizing soil surfaces, and enhancing infiltration rates. Plant roots bind soil particles together, creating a network of soil aggregates that resist erosion and promote soil stability. Dense vegetation canopies intercept rainfall, reduce surface runoff, and dissipate runoff energy, minimizing soil detachment and transport. Riparian vegetation along streambanks and hillslopes provides additional protection against erosion and sedimentation within watersheds.

f. **Soil Properties**: Soil characteristics, including texture, structure, organic matter content, and compaction, influence soil erosion susceptibility within watersheds. Soils with sandy textures and low organic matter content are more prone to erosion due to their low water-holding capacity and poor aggregate stability. Compacted soils have reduced permeability and infiltration rates, leading to increased surface runoff and erosion risk. Soil conservation practices such as cover cropping, contour plowing, terracing, and mulching help protect soil structure and minimize erosion within watersheds.

Impacts of Soil Erosion

Soil erosion within watersheds can have profound impacts on environmental quality, economic productivity, and social well-being, affecting various aspects of watershed health. The impacts of soil erosion on watershed health are multifaceted, spanning environmental, economic, and social dimensions. Addressing soil erosion requires integrated watershed management approaches, soil conservation measures, land use planning strategies, and stakeholder engagement efforts aimed at mitigating erosion risks, restoring degraded landscapes, and promoting sustainable land management practices within watersheds. Here's a discussion of the environmental, economic, and social consequences of soil erosion:

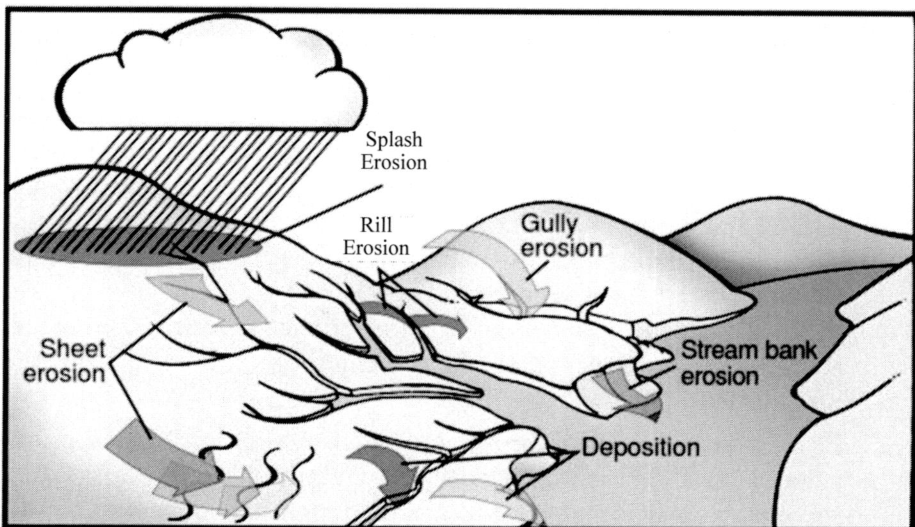

Figure 9. Soil erosion by Daniel D. Richter (2014)

a. **Sedimentation:** Soil erosion contributes to the transport of sediment from upland areas to downstream water bodies, leading to sedimentation in rivers, lakes, reservoirs, and coastal ecosystems. Excessive sedimentation can reduce water storage capacity, impair aquatic habitats, and degrade water quality by increasing turbidity, nutrient loading, and sediment-associated pollutants. Sedimentation impacts aquatic biodiversity, fisheries, and recreational activities, as well as infrastructure such as dams, bridges, and irrigation systems.

b. **Water Quality Degradation:** Soil erosion can result in the loss of topsoil, organic matter, and nutrient-rich soil particles, leading to degradation of water quality within watersheds. Sediment-bound nutrients such as phosphorus and nitrogen can contribute to eutrophication, algal blooms, and oxygen depletion in aquatic ecosystems, affecting water clarity, taste, and odor. Soil erosion may also transport contaminants such as pesticides, heavy metals, and pathogens from agricultural fields, urban areas, and industrial sites into surface water bodies, posing risks to human health and ecosystem integrity.

c. **Loss of Agricultural Productivity:** Soil erosion reduces soil fertility, moisture retention, and nutrient availability, compromising agricultural productivity and food security within watersheds. Eroded soils have diminished capacity to support plant growth, resulting in reduced crop yields, decreased crop quality, and increased susceptibility to drought and pest infestations. Soil erosion can lead to loss of arable land, soil degradation, and desertification, exacerbating land degradation and rural poverty in affected areas.

d. **Economic Costs:** Soil erosion imposes significant economic costs on society through direct and indirect impacts on agriculture, water resources, infrastructure, and ecosystem services. The loss of soil fertility and productivity reduces farm income and agricultural output, affecting livelihoods and rural economies dependent on agriculture. Sedimentation-related damages to infrastructure, such as dredging, sediment removal, and flood control measures, incur substantial maintenance and repair costs for governments, utilities, and private landowners.

e. **Erosion-induced Hazards:** Soil erosion increases the risk of erosion-induced hazards such as landslides, mudslides, and debris flows, particularly in steeply sloped or unstable areas within watersheds. Erosion-prone landscapes may experience mass wasting events during heavy rainfall, earthquakes, or land disturbances, posing threats to human safety, property, and infrastructure. Landslide debris can block roads, rivers, and utility corridors, disrupting transportation networks, water supply systems, and energy distribution networks within watersheds.

f. **Social Impacts:** Soil erosion affects communities through impacts on water availability, food security, livelihoods, and cultural heritage within watersheds. Declining water quality and quantity reduce access to safe drinking water, sanitation, and irrigation for rural and urban populations. Reduced agricultural productivity exacerbates food insecurity, malnutrition, and poverty among vulnerable populations dependent on subsistence farming. Soil erosion can also degrade cultural landscapes, archaeological sites, and indigenous territories, eroding cultural identity and traditional land-use practices within watersheds.

2.5 Vegetation and Biodiversity in Watershed Ecosystems

Vegetation Types and Distribution

The vegetation types and distribution within watersheds are diverse and dynamic, reflecting variations in climate, topography, soil conditions, and human influences across different landscapes. The diversity and distribution of vegetation within watersheds play a critical role in shaping watershed health, ecological resilience, and human well-being. By understanding the characteristics and functions of different vegetation types, stakeholders can develop strategies for sustainable land management, biodiversity conservation, and watershed restoration efforts aimed at promoting ecosystem services and resilience within watersheds. Here's an examination of the various vegetation types commonly found within watersheds:

a. **Forests:** Forests are dominant vegetation types within many watersheds, comprising a variety of tree species adapted to specific environmental conditions. Forest ecosystems provide numerous ecological services, including habitat provision, carbon sequestration, water regulation, and soil stabilization. Watershed forests may include temperate forests, tropical rainforests, montane forests, and riparian forests, each with distinct species compositions, canopy structures, and ecological functions.

b. **Grasslands:** Grasslands are open ecosystems dominated by herbaceous plants such as grasses, sedges, and forbs, adapted to frequent disturbances such as grazing, fire, and drought. Grasslands occur in a range of climatic zones, from temperate prairies and savannas to arid steppes and tundras. Grassland ecosystems support diverse wildlife, including grazing mammals, birds, and insects, and provide valuable forage for livestock grazing within watersheds.

c. **Wetlands:** Wetlands are transitional ecosystems characterized by saturated soils, standing water, and unique hydrological regimes, supporting a diverse array of aquatic and terrestrial vegetation. Wetlands include marshes, swamps, bogs, fens, and mangrove forests, which provide critical functions such as flood mitigation, water filtration, nutrient cycling, and wildlife habitat within watersheds. Wetlands serve as nurseries for fish and wildlife, breeding grounds for migratory birds, and buffers against shoreline erosion and storm surge.

d. **Riparian Zones:** Riparian zones are vegetated corridors along rivers, streams, lakes, and wetlands, characterized by a mix of aquatic and terrestrial plant species adapted to fluctuating water levels and periodic flooding. Riparian vegetation plays a crucial role in stabilizing streambanks, filtering sediment and pollutants, shading water bodies, and providing habitat for aquatic and terrestrial species within watersheds. Riparian forests, shrubs, grasses, and wetland plants contribute to streambank erosion control, water quality improvement, and biodiversity conservation.

e. **Shrublands:** Shrublands, also known as chaparral, scrublands, or heathlands, are ecosystems dominated by woody shrubs, bushes, and small trees adapted to arid or semi-arid conditions. Shrublands occur in Mediterranean climates, desert regions, coastal areas, and mountain slopes, where they play important roles in soil stabilization, fire resilience, and wildlife habitat within watersheds. Shrublands support a variety of plant and animal species adapted to harsh environmental conditions, including drought-tolerant shrubs, succulents, and small mammals.

f. **Cultivated Lands:** Cultivated lands, including croplands, orchards, vineyards, and pastures, are areas where vegetation has been intentionally modified or replaced by human activities such as agriculture, horticulture, and livestock grazing. Cultivated lands within watersheds provide food, fiber, and livelihoods for human communities, but they can also pose environmental challenges such as soil erosion, nutrient runoff, and pesticide contamination if not managed sustainably.

Role of Vegetation in Watershed Health

Vegetation plays a crucial role in maintaining watershed health by providing a wide range of ecological functions and ecosystem services that contribute to the resilience and functioning of aquatic and terrestrial ecosystems. The ecological functions and ecosystem services provided by vegetation are essential for sustaining watershed health, biodiversity, and human well-being. By recognizing the role of vegetation in regulating water, climate, soil, and biodiversity processes, stakeholders can implement nature-based solutions, land management practices, and conservation strategies to enhance watershed resilience, ecosystem services, and community resilience within watersheds.

Table: This table highlights the multifaceted role of vegetation in supporting watershed health in India. Vegetation plays a crucial role in regulating hydrological processes, conserving biodiversity, mitigating climate change impacts, and providing livelihood support to local communities. Ensuring the conservation and sustainable management of vegetation is essential for maintaining the ecological integrity and resilience of watersheds across India.

S. No.	Aspect	Role of Vegetation
1.	Water Infiltration and Soil Stability	Enhances water infiltration into soil, reducing surface runoff and soil erosion.
		Improves soil structure and stability, reducing the risk of landslides and soil erosion during heavy rainfall.
		Increases soil moisture retention, contributing to groundwater recharge and sustained streamflow.

2.	Riparian Buffers	Provides riparian buffers along water bodies, filtering sediments, nutrients, and pollutants from runoff.
		Helps maintain water quality by trapping and removing pollutants before they enter aquatic ecosystems.
		Offers habitat and food sources for aquatic organisms, supporting biodiversity and ecosystem resilience.
3.	Biodiversity Conservation	Supports diverse plant and animal species, promoting ecosystem resilience and adaptive capacity.
		Provides habitat for native flora and fauna, including endangered and endemic species.
		Enhances ecological connectivity and ecosystem functioning, facilitating nutrient cycling and energy flow.
4.	Climate Regulation	Regulates local climate by moderating temperature extremes and reducing heat island effects.
		Contributes to carbon sequestration, mitigating greenhouse gas emissions and climate change impacts.
		Influences regional precipitation patterns through evapotranspiration and atmospheric moisture recycling.
5.	Livelihood Support	Sustains traditional livelihoods such as agriculture, forestry, and non-timber forest product collection.
		Provides ecosystem services such as fuelwood, fodder, medicinal plants, and raw materials for handicrafts.
		Offers opportunities for eco-tourism and nature-based recreation, generating income and employment opportunities.

a. **Soil Stabilization**: Vegetation roots anchor soil particles, preventing erosion and soil loss from water and wind erosion processes. The dense network of plant roots improves soil structure and stability, reducing the risk of landslides, slope erosion, and sedimentation in water bodies within watersheds. Vegetation cover also protects against soil compaction, surface crusting, and desertification, maintaining soil fertility and productivity for agriculture, forestry, and ecosystem functioning.

b. **Water Retention and Infiltration:** Vegetation intercepts rainfall, reducing the erosive impact of raindrops and promoting infiltration of water into the soil. The canopy of trees, shrubs, and grasses acts as a natural buffer, slowing down surface runoff and enhancing water retention within watersheds. Vegetation cover increases soil porosity, permeability, and water-holding capacity, replenishing groundwater reserves, sustaining baseflow in streams, and reducing the risk of flooding and droughts.

c. **Carbon Sequestration**: Vegetation plays a vital role in sequestering atmospheric carbon dioxide through photosynthesis, converting carbon into organic matter and storing it in biomass and soil. Forests, grasslands, wetlands, and other vegetated ecosystems act as carbon sinks, absorbing and storing carbon over long periods, thus mitigating climate change and reducing greenhouse gas emissions within watersheds. Vegetation management practices such as reforestation, afforestation, and agroforestry can enhance carbon sequestration and contribute to climate change adaptation and mitigation efforts.

d. **Habitat Provision:** Vegetation provides essential habitat and food resources for a diverse range of plant and animal species within watersheds, supporting biodiversity and ecological resilience. Forests, wetlands, riparian zones, and grasslands serve as breeding grounds, nesting sites, and foraging areas for wildlife, including mammals, birds, amphibians, reptiles, and insects. Vegetation diversity and structural complexity enhance habitat heterogeneity, species richness, and ecosystem stability, promoting ecosystem services such as pollination, pest control, and nutrient cycling.

e. **Water Quality Improvement**: Vegetation plays a critical role in filtering pollutants, nutrients, and sediments from surface runoff and groundwater, improving water quality in rivers, lakes, wetlands, and aquifers within watersheds. Riparian vegetation, wetland plants, and vegetated buffers act as natural filters, trapping sediment, absorbing nutrients, and metabolizing contaminants before they reach water bodies. Vegetation cover reduces nutrient runoff from agricultural fields, urban areas, and industrial sites, minimizing eutrophication, algal blooms, and waterborne diseases.

f. **Erosion Control and Flood Mitigation**: Vegetation mitigates soil erosion, reduces runoff velocities, and stabilizes streambanks and slopes, minimizing the risk of erosion-induced hazards such as landslides, mudslides, and floods within watersheds. The dense root systems of trees, shrubs, and grasses reinforce soil structure, preventing mass wasting events and streambank erosion. Vegetation buffers along watercourses, contour planting, and terracing practices help to control runoff, retain sediment, and mitigate downstream flooding, protecting communities, infrastructure, and natural habitats.

Biodiversity Conservation

Biodiversity conservation within watersheds is essential for maintaining ecological resilience, supporting ecosystem functions, and enhancing human well-being. Biodiversity conservation within watersheds is essential for maintaining ecosystem health, sustaining ecosystem services, and safeguarding human well-being in the face of global environmental challenges. By protecting and restoring biodiversity-rich habitats, enhancing ecological connectivity, and promoting sustainable land management practices, stakeholders can ensure the resilience and adaptive capacity of watersheds for current and future generations. Here's an exploration of the importance of biodiversity conservation within watersheds:

a. **Ecological Resilience:** Biodiversity provides the foundation for ecosystem resilience, enabling ecosystems to adapt to changing environmental conditions, natural disturbances, and human impacts. High levels of biodiversity within watersheds increase ecosystem stability, resistance to invasive species, and resilience to stressors such as climate change, habitat loss, and pollution. Biodiverse ecosystems exhibit greater capacity for self-regulation, ecosystem recovery, and long-term sustainability, buffering against environmental degradation and maintaining ecosystem services critical for human survival and well-being.

b. **Ecosystem Functions:** Biodiversity drives key ecosystem functions and processes that support life on Earth, including nutrient cycling, water purification, pollination, pest control, and carbon sequestration. Diverse plant and animal species within watersheds contribute to these ecosystem functions through complex interactions and feedback loops, enhancing ecosystem productivity, stability, and resilience. Biodiverse ecosystems provide essential goods and services that sustain human societies, such as clean water, fertile soils, climate regulation, and recreational opportunities, fostering human health, prosperity, and cultural identity.

c. **Habitat Provision**: Biodiversity conservation protects and restores habitat diversity within watersheds, ensuring the survival of diverse plant and animal species adapted to specific ecological niches and microhabitats. Healthy habitats, including forests, wetlands, grasslands, and riparian zones, provide essential resources such as food, shelter, breeding sites, and migration corridors for wildlife populations, supporting biodiversity hotspots, keystone species, and ecosystem engineers. Protected areas, wildlife reserves, and green corridors connect fragmented habitats, allowing species to disperse, migrate, and adapt to changing environmental conditions, promoting genetic diversity and species resilience within watersheds.

d. **Ecosystem Services**: Biodiversity underpins the delivery of ecosystem services that directly and indirectly benefit human societies, including provisioning services (e.g., food, timber, medicine), regulating services (e.g., climate regulation, water purification), supporting services (e.g., soil formation, nutrient cycling), and cultural services (e.g., recreation, spiritual values). Biodiverse watersheds provide a wide range of ecosystem services essential for human well-being, livelihoods, and cultural practices, contributing to economic prosperity, social equity, and environmental sustainability. Biodiversity conservation enhances the capacity of ecosystems to provide these services over the long term, ensuring the resilience and adaptive capacity of watersheds in the face of global environmental changes and socio-economic challenges.

e. **Climate Adaptation and Mitigation**: Biodiversity conservation within watersheds plays a critical role in climate adaptation and mitigation efforts, helping to sequester carbon, regulate hydrological cycles, and buffer communities against climate-related risks such as floods, droughts, and storms. Biodiverse ecosystems such as forests, wetlands, and mangroves act as carbon sinks, absorbing and storing atmospheric carbon dioxide, thus mitigating climate change impacts and reducing greenhouse gas emissions. Biodiversity-rich landscapes enhance ecosystem resilience to climate variability, supporting ecosystem-based adaptation strategies such as agroforestry, reforestation, and habitat restoration within watersheds.

Table. Table categorizes strategies into adaptation and mitigation, providing examples and brief descriptions for each strategy.

S. No.	Category	Strategy	Description	Examples
1.	**Adaptation**	Infrastructure Improvements	Enhancing infrastructure to withstand climate impacts	Building flood defenses, upgrading drainage systems, constructing resilient buildings
		Agricultural Adaptation	Modifying agricultural practices to cope with changing climate conditions	Developing drought-resistant crops, changing planting schedules, implementing water-efficient irrigation
		Water Resource Management	Managing water resources to ensure availability during periods of scarcity	Rainwater harvesting, desalination, improving water-use efficiency

		Ecosystem Protection	Protecting and restoring natural ecosystems to enhance resilience	Wetland restoration, reforestation, creating marine protected areas
		Health System Strengthening	Preparing healthcare systems for climate-related health challenges	Improving disease surveillance, heatwave response plans, enhancing healthcare infrastructure
		Community-based Adaptation	Empowering communities to develop localized adaptation strategies	Participatory planning, local risk assessments, community education and training
2.	**Mitigation**	Renewable Energy Transition	Shifting from fossil fuels to renewable energy sources	Solar power, wind energy, hydroelectric power, geothermal energy
		Energy Efficiency	Reducing energy consumption through more efficient technologies and practices	LED lighting, energy-efficient appliances, building insulation
		Sustainable Transportation	Promoting low-carbon transportation options	Electric vehicles, public transit, cycling infrastructure, car-sharing programs
		Carbon Sequestration	Capturing and storing carbon dioxide from the atmosphere	Afforestation, reforestation, soil carbon sequestration, direct air capture
		Waste Management	Reducing greenhouse gas emissions from waste	Recycling, composting, methane capture from landfills
		Policy and Regulation	Implementing laws and regulations to reduce emissions	Carbon pricing, emissions trading systems, renewable energy mandates

Important Questions

1. How do natural processes such as infiltration and percolation contribute to groundwater recharge within watersheds, and what are some engineered methods used to enhance groundwater replenishment in areas facing water scarcity?
2. How does soil erosion contribute to sedimentation in water bodies, and what are the environmental consequences of sedimentation within watersheds?
3. What are some of the water quality degradation issues associated with soil erosion, and how do these impacts affect both ecosystems and human health within watersheds?

4. Can you explain the relationship between soil erosion and loss of agricultural productivity within watersheds, and how does this impact food security and rural economies?
5. What economic costs are associated with soil erosion, and how do these costs manifest in terms of infrastructure damage, maintenance, and repair within watersheds?
6. How does soil erosion contribute to erosion-induced hazards, and what measures can be taken to mitigate these hazards and ensure community safety within watersheds?
7. What social impacts are associated with soil erosion, particularly in terms of access to water, food security, livelihoods, and cultural heritage within watersheds?
8. How does vegetation contribute to soil stabilization and erosion control within watersheds, and what are some practical implications of this role for watershed management and conservation efforts?
9. How does biodiversity contribute to ecological resilience within watersheds, and what are some examples of biodiversity conservation strategies that enhance ecosystem stability and resistance to environmental stressors?
10. Can you explain the role of biodiversity in driving key ecosystem functions and services within watersheds, and how do these services support human well-being and sustainable development?
11. What are the implications of habitat provision through biodiversity conservation within watersheds for wildlife populations, ecosystem connectivity, and landscape-scale conservation efforts?
12. How does biodiversity conservation contribute to the delivery of ecosystem services, and what are some economic, social, and environmental benefits associated with maintaining biodiversity within watersheds?
13. In what ways does biodiversity conservation within watersheds contribute to climate adaptation and mitigation efforts, and how can biodiversity-rich ecosystems help communities adapt to and mitigate the impacts of climate change?
14. What are some practical approaches and strategies for promoting biodiversity conservation within watersheds, and how can stakeholders collaborate to address biodiversity loss and enhance watershed resilience?

15. How do shrublands differ from forests in terms of plant composition?
16. What factors make shrublands well-suited for withstanding arid conditions?
17. What are some alternative names for shrublands mentioned in the paragraph?
18. In what types of climates do shrublands typically occur?
19. What types of wildlife are commonly found in shrubland ecosystems?
20. Can you provide a specific example from India where the implementation of vegetation-based erosion control and flood mitigation strategies has been successful in protecting communities and infrastructure, while also enhancing natural habitats within watersheds?
21. How do geological, geomorphological, and hydrological processes contribute to the formation of diverse landforms within watersheds?
22. Why is understanding the characteristics and functions of different landforms important for effective watershed management?
23. Can you provide examples of specific landforms and their roles in regulating water flow, sediment transport, and ecosystem functioning within watersheds?
24. How do the interactions between landforms and water influence natural resource conservation efforts within watersheds?
25. What are some examples of how changes in land cover can impact the infiltration rate and runoff patterns within a watershed?

3

Assessment of Land, Water and Biomass Resources

3.1 Land Use and Land Cover Analysis

Concept and Importance: Understanding the significance of land use and land cover analysis in watershed management for assessing spatial patterns of human activities and natural ecosystems.

Land use and land cover analysis is a fundamental tool in watershed management, providing valuable insights into the spatial patterns of human activities and natural ecosystems within watersheds. Here's an exploration of the concept and importance of land use and land cover analysis in watershed management:

a. **Concept:** Land use refers to the human activities and practices conducted on the land, including residential, commercial, industrial, agricultural, recreational, and conservation activities. Land cover, on the other hand, refers to the physical and biological features of the land surface, including vegetation, water bodies, bare soil, impervious surfaces, and artificial structures. Land use and land cover analysis involves mapping, classifying, and quantifying the spatial distribution and temporal changes of different land use and land cover types within watersheds using remote sensing, Geographic Information Systems (GIS), and field surveys.

b. **Significance:** Land use and land cover analysis is essential for watershed management due to its significance in understanding the interactions between human activities and natural processes, assessing environmental impacts, and guiding land management decisions. Key reasons for the importance of land use and land cover analysis in watershed management include:

c. **Resource Allocation**: Land use and land cover data provide information on the allocation of land resources for different purposes, including agriculture, urban development, conservation, and infrastructure planning. Understanding land use patterns helps stakeholders prioritize land management actions, allocate resources effectively, and balance competing land uses to meet socio-economic and environmental objectives within watersheds.

d. **Environmental Monitoring**: Land use and land cover analysis facilitate the monitoring of environmental changes, including deforestation, urban expansion, agricultural intensification, wetland loss, and habitat fragmentation within watersheds. Monitoring land cover changes over time helps identify trends, drivers, and hotspots of land transformation, enabling early detection of environmental degradation, habitat loss, and ecosystem disturbances.

e. **Hydrological Modeling**: Land use and land cover data are essential inputs for hydrological modeling and watershed modeling studies, as they influence surface runoff, infiltration, evapotranspiration, and water quality processes within watersheds. Land use maps provide information on land surface characteristics, vegetation cover, imperviousness, and soil properties, which are used to parameterize hydrological models and simulate hydrological processes under different land management scenarios.

f. **Ecosystem Services Assessment**: Land use and land cover analysis support the assessment of ecosystem services provided by natural and semi-natural ecosystems within watersheds, including water provisioning, carbon sequestration, biodiversity conservation, and recreational opportunities. Mapping ecosystem services helps quantify their value, prioritize conservation efforts, and integrate ecosystem-based approaches into watershed management strategies to enhance human well-being and ecosystem resilience.

g. **Land Management Planning**: Land use and land cover information inform land management planning and decision-making processes, guiding land use zoning, land use regulations, land acquisition, and restoration efforts within watersheds. Spatial analysis of land cover data helps identify land use conflicts, land degradation hotspots, and areas of high conservation value, facilitating participatory land use planning processes and stakeholder engagement in watershed management initiatives.

In summary, land use and land cover analysis play a critical role in watershed management by providing spatially explicit information on land use dynamics, environmental changes, and ecosystem services within watersheds. By integrating land use and land cover data into decision-making processes, stakeholders can promote sustainable land management practices, protect natural resources, and enhance watershed resilience to global environmental changes and socio-economic pressures.

Methods and Techniques: Exploring various methods and techniques used for land use and land cover classification, including satellite imagery interpretation, GIS analysis, and field surveys. Methods and techniques for land use and land cover classification are diverse, ranging from traditional field surveys to advanced remote sensing and Geographic Information Systems (GIS) analysis. Here's an exploration of various methods and techniques used for land use and land cover classification in watershed management:

a. **Satellite Imagery Interpretation**: Satellite imagery interpretation is a common method for land use and land cover classification, involving the visual analysis of satellite images to identify and classify different land cover types within watersheds. Interpretation relies on spectral signatures, spatial patterns, and contextual information to differentiate between land cover classes such as forests, agricultural fields, urban areas, water bodies, and natural habitats. Visual interpretation can be conducted using various satellite sensors, including optical (e.g., Landsat, Sentinel) and multispectral (e.g., MODIS, SPOT) imagery, at different spatial and temporal resolutions.

b. **Remote Sensing Classification:** Remote sensing classification techniques involve the automated analysis of satellite or aerial imagery to classify land cover types based on spectral, spatial, and temporal characteristics. Remote sensing classification algorithms use mathematical algorithms, machine learning algorithms, and image processing techniques to segment, classify, and map land cover classes within watersheds. Common classification methods include supervised classification, unsupervised classification, object-based image analysis (OBIA), and machine learning techniques such as support vector machines (SVM), random forests, and neural networks.

c. **Geographic Information Systems (GIS) Analysis**: GIS analysis integrates spatial data on land use and land cover with other geographic datasets (e.g., topography, hydrology, soils) to analyze spatial relationships, patterns, and trends within watersheds. GIS-based land use mapping involves overlaying, buffering, spatial querying, and spatial modeling techniques to analyze land cover change, land use suitability, and landscape metrics within watersheds. GIS software packages such as ArcGIS, QGIS, and ENVI are commonly used for land use and land cover analysis, allowing users to visualize, manipulate, and analyze spatial data effectively.

d. **Field Surveys and Ground Truthing:** Field surveys and ground truthing are essential components of land use and land cover classification, providing ground-based validation, calibration, and verification of remote sensing data and classification results. Field surveys involve collecting field data on land cover types, vegetation species, land use practices, and landscape features within watersheds using GPS, field notes, and photographic documentation. Ground truthing helps identify errors, inaccuracies, and discrepancies in remote sensing classification results, improving the accuracy and reliability of land cover maps and spatial datasets.

e. **Object-Based Image Analysis (OBIA):** Object-based image analysis (OBIA) is a spatial analysis approach that segments satellite imagery into meaningful objects or image objects based on spectral, spatial, and contextual properties. OBIA combines spectral information with geometric and contextual information to classify land cover types and extract thematic information from imagery within watersheds. OBIA techniques enable the identification of complex land cover patterns, edge detection, and feature extraction, improving the accuracy and interpretability of land cover classification results for watershed management applications.

f. **Change Detection Analysis:** Change detection analysis compares multi-temporal satellite images to detect and quantify changes in land use and land cover over time within watersheds. Change detection techniques involve image differencing, image rationing, and image classification methods to identify areas of land cover change, land cover conversion, and land cover dynamics within watersheds. Change detection analysis helps assess the impacts of human activities, natural disturbances, and environmental changes on watershed health, informing land management decisions and conservation strategies.

Overall, a combination of methods and techniques, including satellite imagery interpretation, remote sensing classification, GIS analysis, field surveys, OBIA, and change detection analysis, is employed for land use and land cover classification in watershed management. Integrating these methods allows stakeholders to generate accurate, reliable, and up-to-date land cover maps, spatial datasets, and information products for informed decision-making, sustainable land management planning, and conservation actions within watersheds.

- **Land Use Change Detection:** Discussing approaches for detecting and monitoring changes in land use and land cover over time, including urbanization, deforestation, agriculture expansion, and land degradation.

Land use change detection is crucial for understanding the dynamics of human-environment interactions within watersheds, including urbanization, deforestation, agriculture expansion, and land degradation. Here's a discussion of approaches for detecting and monitoring changes in land use and land cover over time:

a. **Remote Sensing Analysis**: Remote sensing analysis is a powerful approach for detecting and monitoring land use changes using satellite imagery acquired at different time intervals. Various techniques, including image differencing, image rationing, and image classification, are employed to compare multi-temporal satellite images and identify areas of land use change within watersheds. Change detection algorithms analyze spectral, spatial, and temporal differences between image pairs to detect changes in land cover types, such as urban growth, forest loss, agricultural expansion, and wetland conversion. Remote sensing analysis allows for the quantification, visualization, and mapping of land use changes over large spatial extents, providing valuable insights into landscape dynamics and drivers of change within watersheds.

b. **Supervised Classification:** Supervised classification is a commonly used method for detecting land use changes by training classification algorithms to differentiate between land cover classes based on spectral signatures. Supervised classifiers, such as maximum likelihood, support vector machines (SVM), and random forests, are trained using reference samples of land cover types from multi-temporal satellite imagery. The trained classifiers are then applied to classify each image into land cover classes and identify changes in land use over time within watersheds. Supervised classification allows for the detection of specific land cover changes, such as urban expansion, agricultural intensification, and forest fragmentation, by comparing classified images from different time periods.

c. **Change Vector Analysis (CVA):** Change vector analysis (CVA) is a change detection technique that quantifies changes in land cover by calculating the magnitude and direction of change vectors between multi-temporal satellite images. CVA compares pixel values between image pairs and computes the difference in spectral values to determine the magnitude of change in each pixel. The direction of change vectors indicates the type of land cover change (e.g., urbanization, deforestation, agriculture) within watersheds. CVA provides a spatially explicit representation of land use changes, allowing for the identification of hotspots, trends, and patterns of change over time.

d. **Post-Classification Comparison**: Post-classification comparison is a method for detecting land use changes by comparing independently classified images from different time periods. Classified images are generated for each time period using supervised classification techniques, and land cover change maps are created by comparing the classification results pixel by pixel. Post-classification comparison allows for the identification of changes in land cover categories (e.g., forest to agriculture, water to urban) within watersheds, providing insights into the drivers and impacts of land use changes over time.

e. **Time Series Analysis:** Time series analysis involves analyzing multi-temporal satellite imagery acquired over a continuous time period to track changes in land use and land cover dynamics within watersheds. Time series analysis techniques, such as vegetation indices, change detection algorithms, and trend analysis, are used to identify trends, seasonality, and anomalies in land cover patterns over time. Time series analysis allows for the detection of gradual changes (e.g., forest degradation, agricultural expansion) and abrupt changes (e.g., urbanization, natural disasters) in land use within watersheds, supporting long-term monitoring and management efforts.

By employing these approaches for land use change detection, stakeholders can assess the extent, magnitude, and drivers of land use changes within watersheds, informing land management decisions, conservation strategies, and sustainable development planning initiatives aimed at preserving ecosystem integrity and supporting human well-being over time.

- **Implications for Watershed Management**: Examining the implications of land use and land cover changes on hydrological processes, soil erosion, biodiversity loss, and ecosystem services within watersheds.

Land use and land cover changes have significant implications for watershed management, affecting hydrological processes, soil erosion, biodiversity loss, and ecosystem services within watersheds. Here's an examination of these implications:

Table: Table provides a concise overview of the implications of land use and land cover changes on various aspects of watershed management in India.

S. No.	Aspect	Implications
1	Hydrological Processes	Alteration of water balance, leading to changes in streamflow patterns and groundwater recharge rates.
		Increased surface runoff and flood risk due to decreased infiltration rates.
		Changes in evapotranspiration rates affecting regional precipitation patterns.
2	Soil Erosion	Increased soil erosion rates due to loss of vegetation cover and exposure of soil to erosive forces.
		Sedimentation of water bodies, reducing water storage capacity and degrading aquatic habitats.
3	Biodiversity Loss	Habitat fragmentation and isolation, leading to loss of species diversity and ecosystem resilience.
		Spread of invasive species, disrupting native ecosystems and reducing habitat suitability.
4	Ecosystem Services	Decline in water quality due to sedimentation and nutrient runoff, impacting human health and aquatic ecosystems.
		Reduction in carbon sequestration capacity, contributing to greenhouse gas emissions and climate change.

a. **Hydrological Processes**: Land use and land cover changes alter the hydrological cycle within watersheds, affecting surface runoff, infiltration, groundwater recharge, and streamflow patterns. Urbanization, deforestation, and agricultural expansion can increase surface runoff and peak flows, leading to higher flood risk, reduced baseflow, and altered flow regimes in rivers and streams. Changes in land cover types, such as impervious surfaces, vegetation loss, and soil compaction, can decrease infiltration rates, increase runoff volumes, and exacerbate erosion and sedimentation in water bodies within watersheds. Land use changes may also impact water quality by introducing pollutants, nutrients, and contaminants into surface water and groundwater systems, affecting aquatic ecosystems and human health.

b. **Soil Erosion:** Land use and land cover changes influence soil erosion rates, sediment transport, and sediment delivery within watersheds, leading to soil degradation, land degradation, and loss of soil fertility. Deforestation, agricultural expansion, and land development activities remove vegetative cover, expose soil to erosive forces, and increase soil erosion risk in vulnerable landscapes. Erosion-prone areas, such as steep slopes, erodible soils, and degraded lands, are particularly susceptible to erosion hazards, resulting in sedimentation in water bodies, sediment pollution, and habitat degradation within watersheds. Soil erosion reduces soil productivity, impairs water quality, and undermines ecosystem resilience, posing challenges for agricultural productivity, water resource management, and watershed restoration efforts.

c. **Biodiversity Loss:** Land use and land cover changes contribute to habitat loss, fragmentation, and degradation, leading to biodiversity loss and ecosystem degradation within watersheds. Habitat conversion, urban sprawl, and infrastructure development fragment natural landscapes, disrupt ecological connectivity, and isolate wildlife populations, reducing species richness, genetic diversity, and ecosystem resilience. Loss of native habitats, such as forests, wetlands, and riparian zones, threatens plant and animal species adapted to specific ecological niches, leading to species extinctions, population declines, and ecological imbalances within watersheds. Biodiversity loss diminishes ecosystem services, impairs ecosystem functions, and compromises the capacity of ecosystems to provide essential goods and services for human well-being and ecosystem health.

d. **Ecosystem Services:** Land use and land cover changes affect the provision of ecosystem services, including water provisioning, carbon sequestration, nutrient cycling, and cultural services, within watersheds. Changes in land cover types, such as forest conversion to agriculture, wetland drainage, and urbanization, alter the capacity of ecosystems to regulate hydrological processes, mitigate climate change, and support biodiversity conservation efforts. Degradation of ecosystem services, such as reduced water availability, degraded water quality, and loss of recreational opportunities, can undermine human livelihoods, community resilience, and socio-economic development within watersheds. Sustainable land use planning, conservation strategies, and watershed management interventions are needed to mitigate the adverse impacts of land use changes and safeguard ecosystem services for present and future generations.

In summary, land use and land cover changes have far-reaching implications for watershed management, posing challenges for hydrological processes, soil erosion control, biodiversity conservation, and ecosystem services provision within watersheds. Addressing these implications requires integrated approaches, stakeholder engagement, and adaptive management strategies aimed at promoting sustainable land use practices, conserving natural resources, and enhancing watershed resilience to global environmental changes and human pressures.

3.2 Water Resources: Inventory and Assessment

Water Resource Components: Identifying and quantifying the different components of water resources within watersheds, including surface water bodies, groundwater aquifers, and precipitation. Water resource components within watersheds encompass various elements that collectively contribute to the availability, distribution, and quality of water within a given area. Here's an identification and quantification of the different components of water resources within watersheds:

a. **Surface Water Bodies**: Surface water bodies constitute one of the primary components of water resources within watersheds. These include rivers, streams, lakes, reservoirs, ponds, and wetlands. Surface water bodies receive water from precipitation, runoff, and groundwater discharge, and they play a vital role in hydrological processes such as water storage, conveyance, and regulation. Surface water bodies serve as important sources of freshwater for various human activities, including drinking water supply, irrigation, industrial use, and recreation. They also support diverse aquatic ecosystems, providing habitat for fish, wildlife, and aquatic plants.

b. **Groundwater Aquifers**: Groundwater aquifers are another essential component of water resources within watersheds. Aquifers are underground layers of permeable rock or sediment that contain water and can supply groundwater to wells, springs, and surface water bodies. Groundwater aquifers store vast amounts of freshwater and serve as a reliable source of drinking water, agricultural irrigation, and industrial use within watersheds. Groundwater recharge occurs through infiltration of precipitation, percolation of surface water, and lateral flow from adjacent areas. Groundwater discharge occurs through springs, seeps, and baseflow to rivers and streams, contributing to streamflow and maintaining ecosystem health.

c. **Precipitation:** Precipitation is a fundamental component of water resources within watersheds, representing the input of water into the hydrological cycle. Precipitation includes rainfall, snowfall, sleet, and hail, and it is the primary source of freshwater for terrestrial ecosystems and human communities within watersheds. Precipitation is distributed unevenly across watersheds due to variations in climate, topography, and atmospheric conditions. It replenishes surface water bodies, recharges groundwater aquifers, and sustains soil moisture levels, supporting plant growth, agricultural productivity, and ecosystem functioning within watersheds. Precipitation patterns influence hydrological processes such as evaporation, transpiration, infiltration, and runoff, shaping watershed hydrology and water availability throughout the year.

Quantifying and understanding the dynamics of these water resource components within watersheds is essential for water resources management, watershed planning, and sustainable development initiatives. By assessing the availability, distribution, and interactions of surface water bodies, groundwater aquifers, and precipitation within watersheds, stakeholders can develop integrated water management strategies, optimize water allocation, and promote water conservation measures to meet the diverse needs of human societies, preserve ecosystem integrity, and enhance watershed resilience to environmental changes and socio-economic pressures.

Hydrological Monitoring: Discussing methods and techniques for monitoring hydrological parameters, such as streamflow, groundwater levels, precipitation patterns, and water quality indicators. Hydrological monitoring plays a crucial role in understanding the dynamics of water resources within watersheds and informing water management decisions. Here's a discussion of methods and techniques for monitoring hydrological parameters, including streamflow, groundwater levels, precipitation patterns, and water quality indicators:

a. **Streamflow Monitoring**: Streamflow monitoring involves measuring the discharge of water flowing through rivers, streams, and channels within watersheds. Various methods are used to measure streamflow, including:

b. **Stream Gauging Stations**: Stream gauging stations are equipped with instruments such as streamflow meters, flumes, weirs, and stage sensors to measure water levels and calculate discharge. Continuous or periodic measurements of water levels are taken to estimate flow rates, velocity, and volumetric flow using rating curves and hydraulic formulas.

c. **Flow Velocity Measurements**: Flow velocity can be measured directly using current meters, acoustic Doppler velocimeters (ADV), or velocity-area methods. Velocity measurements are combined with cross-sectional area measurements to calculate streamflow using the Manning's equation or other velocity-area methods.

d. **Remote Sensing**: Remote sensing techniques, such as satellite imagery and aerial photography, can be used to estimate streamflow indirectly by monitoring changes in water surface elevation, width, and extent over time. Remote sensing data can complement ground-based measurements and provide spatially distributed information on streamflow dynamics at regional and watershed scales.

e. **Groundwater Monitoring**: Groundwater monitoring involves measuring groundwater levels, quality, and flow rates in aquifers within watersheds. Common methods and techniques for groundwater monitoring include:

f. **Monitoring Wells**: Monitoring wells are installed at strategic locations within aquifers to measure groundwater levels using water level loggers, pressure transducers, or manual water level measurements. Continuous or periodic measurements of water levels are taken to monitor groundwater fluctuations over time and assess aquifer responses to recharge, extraction, and seasonal variations.

g. **Piezometers:** Piezometers are specialized monitoring wells designed to measure groundwater pressure, hydraulic head, and water quality parameters such as temperature, pH, conductivity, and dissolved oxygen. Piezometers provide insights into aquifer properties, groundwater flow directions, and contaminant transport processes within watersheds.

h. **Pump Tests**: Pump tests are conducted to assess aquifer characteristics, hydraulic properties, and pumping capacities by measuring changes in groundwater levels in response to pumping or injection of water. Pumping tests provide data on aquifer permeability, transmissivity, storativity, and recharge rates, which are used to estimate groundwater resources and develop groundwater management strategies.

i. **Precipitation Monitoring**: Precipitation monitoring involves measuring rainfall, snowfall, and other forms of precipitation within watersheds to assess water inputs into the hydrological cycle. Common methods and techniques for precipitation monitoring include:

j. **Rain Gauges**: Rain gauges are instruments used to collect and measure rainfall amounts at specific locations within watersheds. Standard rain gauges, tipping bucket rain gauges, and weighing rain gauges are commonly used to measure rainfall intensity, duration, and cumulative precipitation over time.

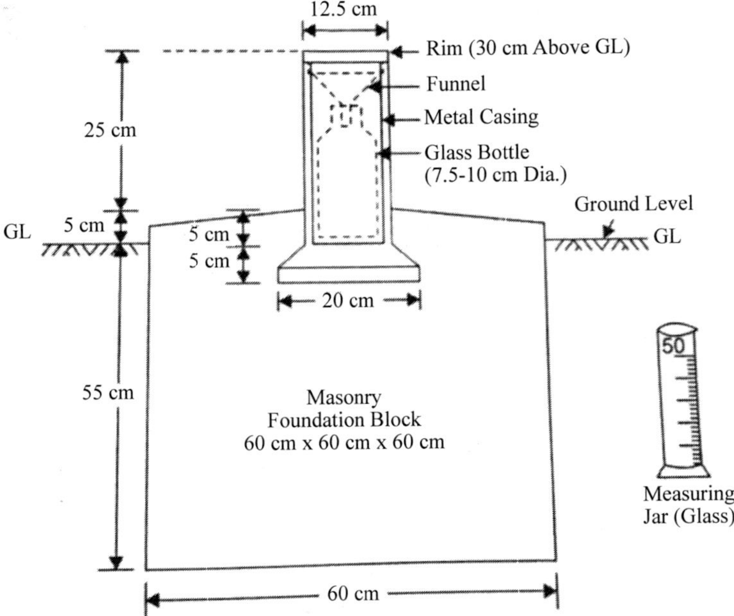

Figure 10. Symon's rain gauge

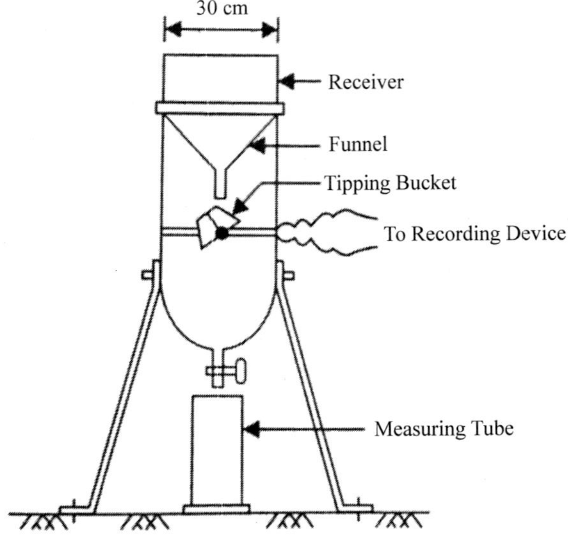

Figure 11. Tipping bucket type rain gauge

k. **Weather Stations**: Weather stations are equipped with multiple sensors to measure meteorological parameters such as temperature, humidity, wind speed, wind direction, and barometric pressure. Weather stations provide comprehensive data on weather conditions, climate patterns, and atmospheric variables that influence precipitation patterns within watersheds.

l. **Radar and Satellite Imagery**: Radar and satellite imagery can be used to estimate precipitation amounts and distribution over large spatial scales by detecting cloud patterns, precipitation rates, and storm trajectories. Radar-based rainfall estimates and satellite-derived precipitation data provide valuable information for hydrological modeling, flood forecasting, and water resources management within watersheds.

m. **Water Quality Monitoring:** Water quality monitoring involves measuring physical, chemical, and biological parameters of surface water and groundwater within watersheds to assess water quality conditions, identify pollution sources, and evaluate ecological health. Common methods and techniques for water quality monitoring include:

n. **Grab Sampling**: Grab sampling involves collecting water samples at specific locations and depths within water bodies using sampling bottles or automated samplers. Water samples are analyzed for various water quality parameters, including pH, temperature, dissolved oxygen, turbidity, conductivity, nutrients (e.g., nitrogen, phosphorus), heavy metals, and organic pollutants.

o. ***In-situ* Sensors**: In-situ sensors and probes are deployed in water bodies to measure real-time water quality parameters continuously. In-situ sensors can measure parameters such as temperature, dissolved oxygen, pH, conductivity, and turbidity at high temporal resolutions, providing insights into diurnal variations and short-term fluctuations in water quality within watersheds.

p. **Biological Monitoring**: Biological monitoring involves assessing the abundance, diversity, and health of aquatic organisms (e.g., macroinvertebrates, fish, algae) as indicators of water quality and ecosystem condition within watersheds. Biological monitoring techniques include habitat assessments, bioassays, biomonitoring surveys, and ecological indices to evaluate ecosystem health and ecological integrity.

By employing these methods and techniques for hydrological monitoring, stakeholders can collect comprehensive data on streamflow, groundwater levels, precipitation patterns, and water quality indicators within watersheds. These data support informed decision-making, watershed planning, and adaptive management strategies aimed at optimizing water resources, protecting aquatic ecosystems, and ensuring sustainable water use for present and future generations.

Water Availability and Demand: Assessing the availability and demand for water resources within watersheds, considering factors such as population growth, land use changes, climate variability, and water use practices. Assessing water availability and demand within watersheds is essential for sustainable water resources management and planning. Here's an overview of the factors considered when assessing water availability and demand:

a. **Population Growth:** Population growth is a significant factor influencing water demand within watersheds. As populations increase, so does the demand for water for domestic, industrial, and agricultural purposes. Population growth projections are used to estimate future water demand and plan for adequate water supply infrastructure, conservation measures, and allocation strategies within watersheds.

b. **Land Use Changes**: Changes in land use patterns, such as urbanization, deforestation, and agricultural expansion, can impact water availability and demand within watersheds. Urbanization leads to increased impervious surfaces, reduced infiltration rates, and higher stormwater runoff, altering hydrological processes and increasing flood risk in urban areas. Agricultural expansion consumes large quantities of water for irrigation, affecting surface water and groundwater resources. Assessing land use changes helps anticipate changes in water demand and manage competing land uses to balance water availability within watersheds.

c. **Climate Variability:** Climate variability, including changes in precipitation patterns, temperature extremes, and hydrological cycles, affects water availability and demand within watersheds. Shifts in precipitation regimes, such as changes in rainfall intensity, frequency, and distribution, influence water supply, runoff, and recharge rates. Temperature changes affect evaporation rates, snowmelt timing, and water storage capacities, impacting water availability for ecosystems and human uses. Assessing climate variability helps identify potential water shortages, drought risks, and adaptive measures to mitigate climate-related impacts on water resources within watersheds.

d. **Water Use Practices**: Water use practices, including domestic, industrial, agricultural, and environmental water uses, influence water demand within watersheds. Assessing water use practices involves quantifying water withdrawals, consumption, and return flows for different sectors and activities. Water demand projections consider changes in water use efficiency, technology adoption, and management practices to optimize water allocation and minimize water stress within watersheds.

e. **Water Infrastructure**: Water infrastructure, such as reservoirs, dams, pipelines, and water treatment facilities, plays a critical role in water availability and demand management within watersheds. Assessing water infrastructure involves evaluating storage capacities, conveyance systems, treatment capacities, and distribution networks to meet current and future water demand. Infrastructure upgrades, maintenance, and investments are needed to enhance water supply reliability, improve water quality, and mitigate water scarcity risks within watersheds.

f. **Water Governance and Policy**: Water governance frameworks, regulations, and policies influence water availability and demand management within watersheds. Assessing water governance involves analyzing water rights, allocation mechanisms, pricing strategies, and institutional arrangements governing water resources management. Effective water governance promotes stakeholder engagement, participatory decision-making, and sustainable water use practices to address competing demands and ensure equitable access to water resources within watersheds.

By assessing water availability and demand considering these factors, stakeholders can develop integrated water resources management plans, implement water conservation measures, and allocate water resources effectively to meet the needs of both human societies and ecosystems within watersheds. Sustainable water management practices promote water security, resilience, and sustainability in the face of changing environmental conditions and socio-economic pressures within watersheds.

Water Balance Analysis: Conducting water balance assessments to understand the inflows, outflows, and storage dynamics of water within watersheds and to inform water resource management decisions. Water balance analysis is a fundamental tool for assessing the hydrological processes and water resource dynamics within watersheds. Here's an overview of conducting water balance assessments to understand the inflows, outflows, and storage dynamics of water within watersheds and to inform water resource management decisions:

a. **Inflows**: Inflows into a watershed represent the sources of water that contribute to the total water availability within the basin. Inflows typically include precipitation (P), which can fall as rainfall or snowfall, and surface water inflows from upstream catchments or tributaries. Precipitation data are collected from meteorological stations, weather satellites, or radar systems and are converted to runoff using runoff coefficients or hydrological models. Surface water inflows are measured using stream gauges or flow meters installed at key locations within the watershed.

b. **Outflows:** Outflows from a watershed represent the pathways through which water leaves the basin. Outflows include evapotranspiration (ET), which is the combined loss of water through evaporation from soil and water surfaces and transpiration from plants. Evapotranspiration rates are estimated based on climatic factors, vegetation cover, soil properties, and land use practices. Other outflows include surface runoff (Q), which is the portion of precipitation that flows overland into streams and rivers, and groundwater discharge (Qg), which is the flow of water from aquifers to surface water bodies or springs. Surface runoff is calculated using hydrological models or measured directly using stream gauges, while groundwater discharge is estimated based on groundwater level measurements, aquifer properties, and hydraulic gradients.

c. **Storage Dynamics**: Water storage within a watershed represents the amount of water stored in various reservoirs, including surface water bodies (e.g., lakes, reservoirs) and groundwater aquifers. Storage dynamics are influenced by changes in precipitation, evapotranspiration, runoff, and groundwater recharge rates over time. Changes in water storage are quantified using water balance equations, which account for the difference between inflows and outflows during a specific time period. Positive storage indicates water accumulation, while negative storage indicates water depletion or withdrawal from storage reservoirs. Water storage data are collected using water level measurements, bathymetric surveys, groundwater level measurements, and remote sensing techniques.

d. **Water Balance Equations**: Water balance equations are used to quantify the components of the water balance within a watershed and assess the overall water availability and sustainability. The general form of the water balance equation is:

$$P - ET - Q - Qg = \Delta S$$

Where,

P = Precipitation

ET = Evapotranspiration

Q = Surface runoff

Q g = Groundwater discharge

ΔS = Change in water storage

By calculating the difference between inflows and outflows, the change in water storage over a specific time period can be determined, providing insights into water availability, water use efficiency, and water resource management priorities within the watershed.

e. **Management Implications**: Water balance analysis provides valuable information for water resource management decisions, including water allocation, infrastructure planning, drought preparedness, and ecosystem conservation. By understanding the inflows, outflows, and storage dynamics of water within watersheds, stakeholders can identify water stress hotspots, prioritize water conservation measures, and develop sustainable water management strategies to meet the needs of both human societies and ecosystems while preserving water resources for future generations.

Water balance analysis is a key tool for assessing the hydrological processes, water availability, and sustainability of water resources within watersheds, guiding informed decision-making and adaptive management approaches to address water challenges and promote water security and resilience in the face of changing environmental conditions and socio-economic pressures.

3.3 Biomass and Carbon Stock Assessment

Biomass Measurement Techniques: Exploring methods for quantifying biomass and carbon stocks within watershed ecosystems, including field measurements, remote sensing, and modeling approaches. Exploring methods for quantifying biomass and carbon stocks within watershed ecosystems involves various techniques that encompass field measurements, remote sensing, and modeling approaches. Here's an overview of each:

a. **Field Measurements:**

Harvesting Methods: Traditional field measurements involve harvesting plant material from sample plots within the watershed. Different plant components, such as aboveground biomass (stems, leaves, and branches) and belowground biomass (roots), are separated, dried, and weighed to determine biomass. This method provides accurate biomass estimates but can be labor-intensive and destructive to vegetation.

Allometric Equations: Allometric equations relate easily measurable parameters, such as tree diameter, height, or crown dimensions, to biomass using statistical relationships derived from field data. These equations enable non-destructive estimation of biomass and are particularly useful for large-scale assessments of forest biomass.

Point Sampling: Point sampling involves collecting vegetation data at randomly selected points or transects within the watershed. Measurements may include tree diameter, tree height, species composition, and canopy cover, which are then used to estimate biomass using conversion factors or species-specific equations.

b. **Remote Sensing**

- **Satellite Imagery:** Satellite imagery provides spatially continuous data on land cover, vegetation indices (e.g., NDVI - Normalized Difference Vegetation Index), and canopy structure, which can be used to estimate biomass and carbon stocks within watersheds. Remote sensing techniques, such as spectral unmixing and vegetation indices, allow for the derivation of biomass maps at various spatial resolutions.
- **Lidar (Light Detection and Ranging):** Lidar technology uses laser pulses to measure the three-dimensional structure of vegetation, including canopy height, canopy density, and vertical biomass distribution. Lidar data are valuable for high-resolution mapping of biomass and carbon stocks within forests and other vegetation types in watersheds.

Modeling Approaches

- **Biogeochemical Models:** Biogeochemical models simulate the processes of carbon cycling and biomass accumulation within watershed ecosystems based on input data such as climate, soil properties, vegetation type, and land use. These models integrate information on photosynthesis, respiration, decomposition, and nutrient cycling to estimate biomass and carbon stocks over time.
- **Ecosystem Models**: Ecosystem models simulate the interactions between biotic and abiotic factors within watersheds to predict biomass dynamics and carbon sequestration potential. These models consider factors such as plant growth, soil carbon storage, disturbance events, and land management practices to assess the impacts of environmental changes on biomass and carbon stocks.
- Combining these techniques allows for comprehensive assessments of biomass and carbon stocks within watershed ecosystems, enabling researchers and land managers to monitor changes over time, evaluate the effectiveness of conservation measures, and inform policy decisions related to climate change mitigation and ecosystem management. Each method has its strengths and limitations, and the choice of approach

depends on factors such as the scale of analysis, available resources, and specific research objectives.

- **Carbon Sequestration Potential**: Assessing the capacity of watershed ecosystems to sequester carbon through vegetation growth, soil organic matter accumulation, and land management practices. Assessing the carbon sequestration potential of watershed ecosystems involves evaluating the capacity of these ecosystems to capture and store carbon dioxide (CO2) from the atmosphere through various processes, including vegetation growth, soil organic matter accumulation, and land management practices. Here's an overview of how this assessment can be conducted:

a. Vegetation Growth

Biomass Accumulation: Watershed ecosystems, particularly forests, grasslands, and wetlands, act as carbon sinks by sequestering atmospheric carbon through photosynthesis. Assessing the carbon sequestration potential of vegetation involves quantifying the amount of carbon stored in aboveground biomass (e.g., trees, shrubs) and belowground biomass (e.g., roots) using field measurements or remote sensing techniques.

Net Primary Productivity (NPP): NPP represents the rate at which plants assimilate carbon through photosynthesis and convert it into organic matter. Monitoring NPP within watershed ecosystems provides insights into the carbon sequestration rates and productivity of different vegetation types, which are influenced by factors such as climate, soil fertility, and land use history.

b. Soil Organic Matter Accumulation

Soil Carbon Stocks: Soils play a crucial role in carbon sequestration, acting as a long-term reservoir for organic carbon derived from decomposing plant material and microbial activity. Assessing soil carbon stocks within watershed ecosystems involves sampling soils at different depths and land cover types and analyzing them for organic carbon content using laboratory techniques such as dry combustion or wet oxidation.

Land Use and Management Practices: Land management practices, such as afforestation, reforestation, agroforestry, cover cropping, and reduced tillage, can enhance soil carbon sequestration by promoting organic matter accumulation, reducing soil erosion, and minimizing carbon losses from land disturbance activities.

c. **Carbon Accounting and Modeling**

Carbon Inventories: Carbon inventories are conducted to quantify the amount of carbon stored in vegetation, soils, and other ecosystem components within watershed ecosystems. Inventories may involve field measurements, remote sensing data analysis, and modeling approaches to estimate carbon stocks and changes over time.

Ecosystem Models: Ecosystem models simulate carbon cycling processes within watershed ecosystems, including carbon fluxes (e.g., photosynthesis, respiration, decomposition), carbon storage dynamics, and responses to environmental changes. These models predict the carbon sequestration potential of different land cover types, management scenarios, and climate conditions, informing decision-making and policy development for carbon management and climate change mitigation strategies.

Assessing the carbon sequestration potential of watershed ecosystems requires integrated approaches that consider the interactions between vegetation, soils, hydrology, and land management practices. By quantifying carbon stocks, fluxes, and sequestration rates, stakeholders can identify opportunities to enhance carbon storage, promote ecosystem resilience, and mitigate climate change impacts within watersheds.

- **Carbon Accounting and Monitoring:** Discussing approaches for accounting for changes in biomass and carbon stocks over time and monitoring carbon fluxes within watersheds. Carbon accounting and monitoring play essential roles in tracking changes in biomass and carbon stocks over time and assessing carbon fluxes within watersheds. Here's a discussion of approaches for carbon accounting and monitoring:

a. **Carbon Accounting**

Baseline Inventories: Establishing baseline inventories involves quantifying the initial amount of carbon stored in vegetation, soils, and other ecosystem components within the watershed. This involves conducting field measurements, remote sensing analysis, and data synthesis to estimate carbon stocks at a specific point in time.

Carbon Stock Changes: Monitoring changes in carbon stocks over time involves repeated measurements or assessments of biomass, soil organic carbon, and other carbon pools within the watershed. Changes in carbon stocks can result from natural processes (e.g., vegetation growth, decomposition) or human activities (e.g., deforestation, afforestation), and accounting for these changes provides insights into the effectiveness of carbon management strategies.

Carbon Offsets: Carbon accounting may also involve quantifying carbon offsets or emissions reductions achieved through carbon sequestration projects or land management practices within the watershed. Carbon offset projects aim to mitigate greenhouse gas emissions by investing in activities that enhance carbon sinks or reduce carbon emissions, such as reforestation, afforestation, soil conservation, or renewable energy projects.

b. **Carbon Monitoring**

Field Measurements: Field-based monitoring involves collecting data on vegetation biomass, soil carbon, and other carbon pools within sample plots or transects distributed across the watershed. Field measurements may include tree diameter, height, biomass, soil sampling, and carbon content analysis using laboratory techniques. These data provide direct measurements of carbon stocks and changes over time.

Remote Sensing: Remote sensing techniques, such as satellite imagery and aerial photography, enable spatially explicit monitoring of land cover changes, vegetation dynamics, and carbon stocks within watersheds. Remote sensing data can be used to derive vegetation indices, land cover maps, and biomass estimates, which are valuable for large-scale assessments and trend analysis.

Carbon Flux Measurements: Carbon flux measurements involve quantifying the exchange of carbon dioxide (CO2) between the atmosphere and terrestrial ecosystems within the watershed. Flux tower networks, eddy covariance systems, and chamber-based measurements are used to monitor carbon fluxes in forests, grasslands, wetlands, and agricultural fields. These measurements provide insights into ecosystem carbon dynamics, seasonal variations, and responses to environmental changes.

c. **Integrated Modeling**: Integrated modeling approaches combine field measurements, remote sensing data, and process-based models to simulate carbon cycling processes, predict carbon stocks and fluxes, and assess the impacts of land use, climate change, and management practices on watershed carbon dynamics. Ecosystem models, biogeochemical models, and land surface models are used to simulate carbon sequestration potential, evaluate management scenarios, and inform decision-making for carbon management and climate change mitigation strategies.

By employing these approaches for carbon accounting and monitoring, stakeholders can track changes in biomass and carbon stocks over time, assess the effectiveness of carbon management interventions, and identify opportunities to enhance carbon sequestration and mitigate climate change impacts within watersheds. Continuous monitoring and periodic assessments are essential for adaptive management and informed decision-making to achieve carbon neutrality, resilience, and sustainability in watershed ecosystems.

Implications for Climate Change Mitigation: Examining the role of watershed management in contributing to climate change mitigation efforts through carbon sequestration and ecosystem restoration. Examining the role of watershed management in contributing to climate change mitigation efforts through carbon sequestration and ecosystem restoration reveals significant implications for addressing global climate challenges. Here's an overview:

a. **Carbon Sequestration Potential:** Watershed management practices, such as reforestation, afforestation, agroforestry, and soil conservation, have the potential to sequester large amounts of carbon dioxide (CO2) from the atmosphere. Forest ecosystems, in particular, act as significant carbon sinks, storing carbon in vegetation and soil organic matter. By promoting the restoration and sustainable management of forests, wetlands, grasslands, and other ecosystems within watersheds, carbon sequestration rates can be enhanced, mitigating the accumulation of greenhouse gases in the atmosphere and helping to stabilize global climate patterns.

b. **Ecosystem Restoration**: Ecosystem restoration initiatives within watersheds contribute to climate change mitigation by enhancing carbon stocks, biodiversity, and ecosystem resilience. Restoring degraded habitats, such as degraded forests, degraded wetlands, and degraded grasslands, increases vegetation cover, biomass accumulation, and soil organic carbon storage, thereby sequestering carbon and enhancing ecosystem services such as water purification, flood regulation, and habitat provision. Ecosystem restoration projects also provide opportunities for community engagement, livelihood improvement, and sustainable development, aligning climate change mitigation efforts with broader conservation and socio-economic objectives.

c. **Water Resources Management**: Integrated watershed management approaches that consider climate change impacts, water availability, and ecosystem health are essential for climate change mitigation and adaptation. Sustainable water resources management practices, such as reforestation of riparian buffers, sustainable agriculture practices, and watershed protection measures, help maintain hydrological functions, reduce soil erosion, and enhance water quality, thereby supporting ecosystem resilience and carbon sequestration potential. By safeguarding water resources and ecosystem services, watershed management contributes to climate change resilience and adaptation strategies, reducing vulnerabilities and enhancing the capacity of ecosystems and communities to cope with climate-related risks and uncertainties.

d. **Land-Use Planning and Policy Integration:** Integrating climate change considerations into land-use planning, policy development, and decision-making processes is critical for mainstreaming climate change mitigation and adaptation objectives within watershed management strategies. Land-use policies and regulations that prioritize carbon sequestration, biodiversity conservation, and sustainable land management practices facilitate the implementation of climate-smart approaches that balance environmental conservation, economic development, and social equity goals. By promoting climate-resilient land-use planning and policy integration, watershed management fosters synergies between climate change mitigation, adaptation, and sustainable development objectives, maximizing co-benefits and minimizing trade-offs across different sectors and stakeholders.

Examining the implications of watershed management for climate change mitigation underscores the importance of adopting holistic and integrated approaches that address the complex interrelationships between water resources, ecosystems, and human activities. By harnessing the potential of watershed ecosystems as carbon sinks and promoting ecosystem restoration, sustainable land management, and policy integration, watershed management plays a vital role in advancing climate change mitigation efforts and building resilient communities and ecosystems in the face of global environmental challenges.

3.4 Mapping and Remote Sensing Techniques for Resource Assessment

Remote Sensing Applications: Exploring the use of satellite imagery, aerial photography, and other remote sensing technologies for mapping and monitoring natural resources within watersheds. Exploring the use of remote sensing technologies for mapping and monitoring natural resources within watersheds offers valuable insights into ecosystem dynamics, land cover changes, and environmental trends. Here's an overview of remote sensing applications in watershed management:

a. **Satellite Imagery**

Land Cover Classification: Satellite imagery provides high-resolution data on land cover types, including forests, wetlands, agriculture, urban areas, and water bodies, within watersheds. Supervised and unsupervised classification techniques are used to categorize pixels into land cover classes based on spectral signatures and image characteristics.

Change Detection: Time-series analysis of satellite imagery enables the detection and monitoring of land cover changes, such as deforestation, urban expansion, vegetation dynamics, and ecosystem disturbances, within watersheds. Change detection algorithms identify areas of change by comparing multi-temporal images and quantifying differences in land cover classes over time.

Vegetation Monitoring: Vegetation indices derived from satellite imagery, such as the Normalized Difference Vegetation Index (NDVI), Enhanced Vegetation Index (EVI), and Leaf Area Index (LAI), provide insights into vegetation health, biomass, and productivity within watersheds. Vegetation indices are used to assess vegetation dynamics, monitor drought stress, and estimate carbon sequestration potential.

Water Quality Monitoring: Satellite-based sensors, such as multispectral and hyperspectral imagers, detect water quality parameters, such as water turbidity, chlorophyll-a concentration, and suspended sediment levels, in surface water bodies within watersheds. Remote sensing of water quality facilitates spatial mapping and temporal monitoring of water pollution sources, algal blooms, and aquatic habitat degradation.

b. **Aerial Photography**

High-Resolution Imaging: Aerial photography provides detailed, high-resolution images of watershed landscapes, capturing fine-scale features, such as vegetation structure, landforms, and land use patterns, not visible in satellite imagery. Aerial photography is useful for site-specific assessments, habitat mapping, and infrastructure planning within watersheds.

Orthophoto Generation: Aerial photographs are used to generate orthophotos, which are geometrically corrected images that remove distortions caused by terrain relief and camera perspective. Orthophotos serve as base maps for watershed mapping, land use planning, and environmental monitoring applications.

c. **LiDAR (Light Detection and Ranging)**

Topographic Mapping: LiDAR technology provides high-resolution elevation data and 3D terrain models of watershed landscapes, facilitating accurate topographic mapping, watershed delineation, and hydrological modeling. LiDAR-derived terrain metrics, such as slope, aspect, and drainage patterns, are used to characterize watershed geomorphology and analyze hydrological processes.

Vegetation Structure Analysis: LiDAR data capture detailed information on vegetation height, canopy structure, and biomass distribution within watersheds. LiDAR-derived metrics, such as canopy height models (CHMs) and canopy density maps, quantify vegetation structure and biomass, supporting forest inventory assessments, habitat mapping, and carbon accounting studies.

d. Integration with Geographic Information Systems (GIS)

Spatial Analysis: Remote sensing data are integrated with geographic information systems (GIS) to perform spatial analysis, modeling, and visualization of watershed features and processes. GIS-based tools and techniques, such as spatial interpolation, buffer analysis, and suitability mapping, support watershed management activities, such as land use planning, habitat assessment, and environmental impact analysis.

Decision Support Systems: GIS-based decision support systems (DSS) combine remote sensing data, spatial analysis tools, and decision-making frameworks to support informed decision-making in watershed management. DSS platforms enable stakeholders to analyze complex spatial data, evaluate management alternatives, and prioritize actions for sustainable resource management within watersheds.

By leveraging remote sensing technologies for mapping and monitoring natural resources within watersheds, stakeholders can obtain timely, cost-effective, and spatially explicit information for informed decision-making, sustainable land management, and ecosystem conservation initiatives. Remote sensing applications enhance our understanding of watershed dynamics, support environmental monitoring efforts, and facilitate the implementation of adaptive management strategies to address emerging challenges and promote resilience in watershed ecosystems.

Spatial Analysis Tools: Discussing GIS-based spatial analysis tools and techniques for integrating remote sensing data with ancillary information to derive valuable insights for resource assessment. Discussing GIS-based spatial analysis tools and techniques for integrating remote sensing data with ancillary information offers valuable insights for resource assessment within watersheds. Here's an overview of some commonly used spatial analysis tools and techniques:

a. Image Classification

Supervised Classification: Supervised classification techniques, such as Maximum Likelihood Classification and Support Vector Machines (SVM), are used to classify remote sensing imagery into thematic land cover classes based on spectral signatures and training samples. Supervised classification requires a set of training data representing each land cover class to train the classification algorithm.

Unsupervised Classification: Unsupervised classification techniques, such as K-means clustering and ISODATA, group pixels in remote sensing imagery into clusters based on spectral similarities without prior knowledge of land cover classes. Unsupervised classification is useful for exploratory analysis and identifying patterns in the data.

b. Change Detection

Image Differencing: Image differencing compares multi-temporal remote sensing images pixel by pixel to detect changes in land cover and land use over time. Changes between images are identified based on differences in spectral values, allowing for the identification of areas undergoing land cover change within watersheds.

Normalized Difference Vegetation Index (NDVI) Thresholding: Change detection using NDVI thresholding involves setting thresholds for NDVI values to identify areas of vegetation change, such as deforestation or reforestation, between different time periods. Areas with significant changes in NDVI values indicate alterations in vegetation cover within watersheds.

c. Spatial Interpolation

Kriging: Kriging is a geostatistical interpolation technique used to estimate values at unsampled locations within a spatial dataset based on the spatial autocorrelation of sampled data points. Kriging is commonly used to interpolate environmental variables, such as rainfall, temperature, and soil moisture, within watersheds to create continuous surfaces for resource assessment and modeling purposes.

Inverse Distance Weighting (IDW): IDW is a deterministic interpolation technique that estimates values at unsampled locations based on the inverse distance to nearby sampled points. IDW assumes that values closer to sampled points have a greater influence on the interpolated value, making it suitable for spatial interpolation of point data, such as water quality measurements or soil properties.

d. **Buffer Analysis**

Buffer Creation: Buffer analysis involves creating buffers around specific features, such as streams, wetlands, or land use zones, within watersheds to assess spatial relationships and proximity to other features. Buffers are often used to delineate riparian zones, assess habitat connectivity, and identify areas vulnerable to environmental impacts.

Overlay Analysis: Overlaying buffers with other spatial datasets, such as land cover, land use, or environmental sensitivity maps, allows for the identification of areas of overlap or conflict between different features within watersheds. Overlay analysis helps prioritize conservation efforts, land management practices, and development activities based on spatial constraints and compatibility.

e. **Terrain Analysis**

Digital Elevation Model (DEM) Analysis: DEM analysis involves processing elevation data to derive terrain attributes, such as slope, aspect, curvature, and flow accumulation, within watersheds. Terrain analysis helps characterize watershed topography, identify landforms, and assess hydrological processes, such as surface runoff and erosion potential.

f. **Suitability Modeling**

Multi-Criteria Evaluation (MCE): MCE techniques integrate multiple spatial datasets and criteria to assess the suitability of different locations or land use options within watersheds. MCE combines GIS-based analysis, weighted overlay, and decision rules to generate suitability maps for resource management, conservation planning, and land use allocation decisions.

By employing these GIS-based spatial analysis tools and techniques, stakeholders can integrate remote sensing data with ancillary information to derive valuable insights for resource assessment, environmental monitoring, and decision-making within watersheds. Spatial analysis facilitates the identification of spatial patterns, trends, and relationships between landscape features, supporting sustainable management practices and informed policy development for watershed conservation and ecosystem stewardship.

Multi-temporal Analysis: Conducting multi-temporal analysis of satellite imagery to detect land cover changes, vegetation dynamics, and hydrological variations within watersheds. The methodology for conducting multi-temporal analysis of satellite imagery to detect land cover changes, vegetation dynamics, and hydrological variations within watersheds involves the following steps:

a. **Data Acquisition:** Obtain multi-temporal satellite imagery covering the study area from reliable sources such as Landsat, Sentinel, or MODIS. Ensure that the imagery spans multiple time periods to capture temporal changes in land cover and vegetation dynamics.

b. **Pre-processing:** Pre-process the satellite imagery to enhance its quality and suitability for analysis. This may include radiometric and atmospheric corrections, geometric rectification, and mosaicking to create seamless and consistent image composites.

c. **Image Registration:** Register all the multi-temporal images to a common reference system to ensure spatial alignment and accurate comparison between different time periods. Use ground control points or automated registration techniques for precise alignment.

d. **Change Detection Techniques**: **Pixel-based Change Detection:** Utilize pixel-based change detection methods to compare spectral characteristics of corresponding pixels in different images. Techniques such as image differencing or band ratioing can highlight areas of significant change.

   Object-based Change Detection: Implement object-based change detection approaches to analyze changes at a more detailed level by considering spatial patterns and contextual information. This involves segmenting the images into objects or homogeneous regions and comparing object attributes between different time periods.

e. **Land Cover Classification:** Perform land cover classification separately for each time period using supervised or unsupervised classification algorithms. Classify the imagery into relevant land cover classes such as forests, water bodies, agricultural areas, and urban zones.

   Ensure consistency in classification methodologies and input parameters across different time periods to facilitate accurate change detection analysis.

f. **Vegetation Dynamics Analysis:** Calculate vegetation indices such as Normalized Difference Vegetation Index (NDVI) or Enhanced Vegetation Index (EVI) for each time period to quantify vegetation dynamics and health.

   Analyze temporal trends in vegetation indices to identify areas of vegetation growth, decline, or stress over time, which can provide insights into ecosystem productivity and health.

f. **Hydrological Variations Assessment:** Use satellite imagery and terrain data to analyze hydrological variations within watersheds. This may involve delineating water bodies, mapping changes in water extent, and assessing streamflow dynamics over time.

Incorporate hydrological modeling or spatial analysis techniques to identify factors influencing hydrological variations, such as land cover changes, terrain characteristics, and precipitation patterns.

g. **Integration and Interpretation:** Integrate the results of land cover classification, change detection, vegetation dynamics analysis, and hydrological variations assessment into a comprehensive GIS database.

   Interpret the findings to identify areas of land cover change, vegetation dynamics, and hydrological variability within watersheds. Assess the implications of these changes for ecosystem health, water resources management, and land use planning.

h. **Validation and Accuracy Assessment:** Validate the accuracy of change detection results and land cover classification using ground truth data, validation samples, or independent sources of information.

   Assess the overall accuracy of the analysis and quantify error rates to ensure the reliability of the findings.

i. **Reporting and Visualization:** Prepare maps, figures, and reports summarizing the results of the multi-temporal analysis. Use visualizations to communicate key findings and trends effectively to stakeholders and decision-makers.

   Provide recommendations for conservation measures, land management strategies, and monitoring priorities based on the analysis outcomes.

   By following this methodology, researchers can effectively conduct multi-temporal analysis of satellite imagery to monitor land cover changes, vegetation dynamics, and hydrological variations within watersheds, providing valuable insights for ecosystem management and conservation efforts.

Accuracy Assessment: Evaluating the accuracy and reliability of remote sensing products and derived datasets for resource assessment purposes, including ground truthing and validation techniques. The accuracy assessment for evaluating the reliability of remote sensing products and derived datasets for resource assessment purposes, including ground truthing and validation techniques, involves the following steps:

a. **Ground Truth Data Collection:** Establish ground truth data points or reference samples representing the true conditions of the features or phenomena of interest within the study area.

Collect ground truth data through field surveys, site visits, GPS measurements, or existing datasets from reliable sources such as aerial photography, cadastral maps, or previous studies.

b. **Sampling Design:** Design an appropriate sampling strategy to ensure representative and unbiased selection of ground truth samples across the study area.

   Determine the sample size, spatial distribution, and sampling intensity based on the variability of the features being assessed and the desired level of confidence.

c. **Comparison with Remote Sensing Products:** Compare the remote sensing products or derived datasets (e.g., land cover maps, vegetation indices) with the ground truth data to assess their accuracy and agreement.

   Overlay the remote sensing products with the ground truth samples to identify areas of agreement or disagreement and quantify the extent of discrepancies.

d. **Error Matrix Generation:** Construct an error matrix (confusion matrix) to tabulate the agreement between the remote sensing products and ground truth data.

   Classify the sample units into true positives, false positives, true negatives, and false negatives to calculate accuracy metrics such as overall accuracy, producer's accuracy, user's accuracy, and kappa coefficient.

e. **Accuracy Metrics Calculation:** Calculate accuracy metrics to quantify the performance of the remote sensing products in accurately representing the features of interest.

   Overall Accuracy: The proportion of correctly classified samples relative to the total number of samples.

   Calculation: Overall Accuracy = Number of correctly classified samples/ Total number of samples

Indication: This metric represents the proportion of all samples that are correctly classified by the remote sensing product. A higher overall accuracy indicates better performance of the product in correctly classifying the features of interest.

Producer's Accuracy: The probability that a reference sample is correctly classified by the remote sensing product.

Calculation: Producer's Accuracy= Number of correctly classified samples for a class/ Total number of reference samples for that class

Indication: This metric indicates the probability that a reference sample of a particular class is correctly classified by the remote sensing product. It assesses the ability of the product to correctly identify the features of interest as per the reference data.

User's Accuracy: The probability that a pixel classified by the remote sensing product corresponds to the true condition on the ground.

Calculation: User's Accuracy= Number of correctly classified samples for a class/

Total number of samples classified as that class by the product

Indication: This metric represents the probability that a pixel classified by the remote sensing product actually corresponds to the true condition on the ground. It measures the reliability of the product's classifications from the user's perspective.

Kappa Coefficient: A statistical measure of agreement between the remote sensing products and ground truth data, accounting for chance agreement.

The Kappa coefficient is a statistical measure that quantifies the agreement between the remote sensing product classifications and the ground truth data, taking into account the possibility of agreement occurring by chance. Here's a detailed explanation:

$$k = \frac{P_o - P_e}{1 - P_e}$$

Where,

- PoP_oPo (Observed Agreement) is the proportion of observed agreements between the classifications and the ground truth.
- PeP_ePe (Expected Agreement) is the proportion of agreements expected by chance, calculated as:
- $$P_e = \sum_{i=1}^{k} \left(\frac{N_{i+}.N_{+i}}{N^2} \right)$$

Where N_{i+} is the number of reference sample for class i, N_{+i} is the total number of sample classified as class i by the product, and N is the total number of samples.

Importance: Accounts for Chance Agreement: Unlike overall accuracy, the Kappa coefficient accounts for the possibility that some of the agreement between the classification and the reference data might occur by chance. This makes it a more robust measure of agreement.

Range and Interpretation: The Kappa coefficient ranges from -1 to 1, where:

- 1 indicates perfect agreement.
- 0 indicates no agreement beyond chance.
- Negative values indicate agreement worse than chance.

Evaluation of Classification Quality: By accounting for chance, the Kappa coefficient provides a more nuanced evaluation of the classification quality. A high Kappa value suggests that the classification is reliable and not merely a result of random agreement.

f. **Error Sources Analysis:** Identify and analyze sources of errors and uncertainties in the remote sensing products, including classification errors, sensor limitations, data preprocessing artifacts, and thematic misinterpretations.

   Evaluate the impact of error sources on the accuracy and reliability of the derived datasets and assess their implications for resource assessment and decision-making.

g. **Validation Techniques:** Employ various validation techniques to assess the accuracy and robustness of the remote sensing products, such as cross-validation, split-sample validation, and independent validation using additional reference data.

   Validate the remote sensing products across different time periods, spatial extents, or environmental conditions to test their consistency and generalizability.

h. **Reporting and Interpretation:** Document the results of the accuracy assessment, including accuracy metrics, error matrices, and validation techniques employed.

   Interpret the findings to evaluate the suitability of the remote sensing products for resource assessment purposes and identify areas for improvement or refinement in data processing and analysis.

   By conducting an accuracy assessment using ground truthing and validation techniques, researchers can effectively evaluate the reliability and utility of remote sensing products and derived datasets for resource assessment applications, ensuring the credibility and robustness of the analysis outcomes.

Important Questions

1. What are the primary inflows and outflows considered in water balance analysis within watersheds?

2. How are precipitation data collected and utilized in water balance assessments?
3. Describe the factors influencing evapotranspiration rates and how they are estimated in water balance analysis.
4. Explain the significance of groundwater discharge in the context of water balance within watersheds.
5. How are changes in water storage quantified, and what do positive and negative storage values indicate?
6. Can you outline the general form of the water balance equation and its components?
7. Discuss the management implications of water balance analysis for water resource management decisions.
8. How does water balance analysis contribute to identifying water stress hotspots and prioritizing conservation measures?
9. What role does water balance analysis play in promoting water security and resilience within watersheds?
10. How might stakeholders utilize water balance assessments to address challenges related to climate variability and population growth?
11. What are the key steps involved in conducting an accuracy assessment for remote sensing products?
12. How is ground truth data collected, and why is it important for accuracy assessment?
13. Can you explain the concept of an error matrix and its role in quantifying accuracy metrics?
14. What accuracy metrics are commonly used to evaluate the performance of remote sensing products?
15. How do producer's accuracy and user's accuracy differ, and why are they important in accuracy assessment?
16. What are some potential sources of errors and uncertainties in remote sensing products, and how are they addressed in the accuracy assessment?
17. What validation techniques can be employed to assess the reliability of remote sensing products, and how do they differ?
18. Why is it essential to validate remote sensing products across different spatial and temporal conditions?
19. How are the results of an accuracy assessment interpreted and reported to stakeholders?

20. What are some implications of accuracy assessment findings for resource assessment and decision-making processes?
21. How can cross-validation, split-sample validation, and independent validation be effectively implemented to assess the accuracy and robustness of remote sensing products?
22. What are the key considerations when validating remote sensing products across different time periods, spatial extents, or environmental conditions to ensure consistency and generalizability?
23. How are the overall accuracy, producer's accuracy, and user's accuracy calculated for remote sensing products, and what do these metrics indicate about the performance of the products?
24. What is the Kappa coefficient, how is it calculated, and why is it important for evaluating the agreement between remote sensing products and ground truth data?
25. How can remote sensing products be utilized to accurately assess and monitor land, water, and biomass resources, and what are the key accuracy metrics and validation techniques to ensure reliable results?

4

Socio-economic Condition of People

4.1 Demographic Profile of Watershed Communities

- **Population Distribution**: Examining the spatial distribution and density of population within watershed areas, including rural settlements, urban centers, and peri-urban zones. Analyzing population distribution within watershed areas involves understanding how people are distributed across various types of settlements, including rural, urban, and peri-urban zones. Here's a breakdown of each:

  a. **Rural Settlements**: These are typically smaller communities dispersed throughout the watershed area, often engaged in agriculture, forestry, or other primary industries. Population density in rural areas tends to be lower compared to urban centers. The distribution of rural settlements can be influenced by factors such as terrain, availability of natural resources, and historical land use patterns.

  b. **Urban Centers**: Urban areas within watershed boundaries consist of cities and towns with higher population densities and more developed infrastructure. These areas are characterized by a concentration of economic activities, services, and amenities. Population distribution within urban centers can vary widely, with some areas experiencing high population density in downtown cores and others having more spread-out suburbs.

  c. **Peri-urban Zones**: Peri-urban areas lie on the outskirts of urban centers and are transitional zones between rural and urban areas. These areas often experience rapid population growth and development as urbanization spreads outward from cities. Population distribution in peri-urban zones can be dynamic, with a mix of agricultural land, residential developments, and industrial zones.

Analyzing population distribution within watershed areas involves mapping the spatial distribution of settlements, assessing population densities in different types of areas, and understanding the factors driving population trends and dynamics. This information is crucial for land use planning, resource management, and sustainable development within watersheds.

Geographic information systems (GIS), remote sensing, and demographic data are commonly used tools for conducting population distribution studies in watershed areas.

- **Population Characteristics**: Analyzing demographic attributes such as age structure, gender composition, ethnicity, education levels, and household size within watershed communities. Analyzing population characteristics within watershed communities provides valuable insights into the social dynamics and needs of the people living within these areas. Here's a breakdown of key demographic attributes:
  a. **Age Structure**: Understanding the age distribution of the population is essential for planning services and infrastructure. This includes analyzing the proportion of children, working-age adults, and elderly individuals within watershed communities. Demographic pyramids or age-sex pyramids can visualize age structure, highlighting trends such as an aging population or a youth bulge.
  b. **Gender Composition**: Examining the gender balance within watershed communities helps identify potential disparities and inequalities. Gender-specific issues related to health, education, employment, and access to resources may be identified through gender-disaggregated data analysis.
  c. **Ethnicity:** Assessing the ethnic composition of watershed communities provides insights into cultural diversity and social cohesion. Understanding the distribution of different ethnic groups within the population can inform policies and programs aimed at promoting inclusivity and addressing interethnic relations.
  d. **Education Levels:** Analyzing education levels within watershed communities sheds light on human capital development and potential workforce skills. This includes assessing literacy rates, school enrollment rates, attainment levels, and access to educational opportunities across different demographic groups.
  e. **Household Size and Composition**: Studying household characteristics, such as size and composition, helps understand housing needs, family structures, and social support networks within watershed communities. Factors influencing household size may include cultural norms, socioeconomic status, and urbanization trends.

Analyzing these population characteristics requires access to demographic data collected through surveys, censuses, and administrative records.

Statistical analysis techniques, such as demographic profiling and regression modeling, can be employed to identify trends, correlations, and disparities within watershed communities. This information is valuable for designing targeted interventions, allocating resources effectively, and promoting social development and equity within watershed areas.

- **Migration Patterns**: Exploring patterns of internal and external migration, including rural-urban migration, seasonal migration for livelihoods, and displacement due to environmental changes or infrastructure development. Migration patterns within watershed areas are influenced by a variety of factors, including economic opportunities, social dynamics, and environmental conditions. Understanding these patterns is crucial for assessing population dynamics, planning infrastructure, and addressing social and economic challenges. Here are some key aspects of migration within watershed communities:

  a. **Rural-Urban Migration**: Rural-urban migration involves the movement of people from rural areas to urban centers within or adjacent to watershed boundaries. This migration is often driven by the search for better employment opportunities, education, healthcare, and improved living standards. Analyzing rural-urban migration patterns helps in understanding urbanization trends, population growth in urban areas, and the resulting pressure on infrastructure and services.

  b. **Seasonal Migration for Livelihoods**: Seasonal migration is common in many watershed communities, particularly among agricultural and pastoralist populations. People may move temporarily to other areas within or outside the watershed for employment in sectors such as agriculture, construction, or tourism. Understanding the seasonal dynamics of migration is important for planning social support systems, healthcare services, and infrastructure development.

  c. **Displacement due to Environmental Changes**: Environmental factors such as flooding, drought, deforestation, and land degradation can force people to migrate within or across watershed boundaries. This form of migration, often referred to as environmentally induced migration or displacement, can lead to challenges related to resettlement, loss of livelihoods, and conflicts over resources. Analyzing displacement patterns helps in identifying vulnerable populations and implementing measures for adaptation and resilience.

  d. **Displacement due to Infrastructure Development:** Large-scale infrastructure projects, such as dams, roads, and urbanization projects, can result in the displacement of communities living within watershed areas. Understanding the social and economic impacts of infrastructure-induced displacement is essential for ensuring equitable development, mitigating negative consequences, and promoting community participation in decision-making processes.

Analyzing migration patterns within watershed areas requires interdisciplinary approaches that integrate demographic data, spatial analysis techniques, and qualitative research methods. Stakeholder engagement, including local communities, government agencies, and non-governmental organizations, is also critical for developing sustainable solutions to address the drivers and impacts of migration within watershed communities.

- **Population Trends and Projections:** Assessing past trends and projecting future population growth rates within watershed communities to inform planning and development strategies. Assessing population trends and projecting future growth rates within watershed communities is essential for informed decision-making and sustainable development planning. Here's how it can be done:

a. **Data Collection and Analysis:** Begin by gathering historical demographic data for the watershed communities, including population counts, age distributions, migration patterns, and fertility rates. Analyze these data to identify past trends in population growth, urbanization, and demographic changes.

b. **Demographic Modeling:** Use demographic modeling techniques to project future population trends within the watershed area. This involves applying mathematical models, such as cohort-component models or population projection software, to estimate future population size, age structure, and spatial distribution based on assumptions about fertility, mortality, and migration rates.

c. **Scenario Analysis:** Conduct scenario analysis to explore different potential futures based on varying assumptions about key demographic variables, such as fertility rates, mortality rates, and migration flows. This helps decision-makers understand the range of possible outcomes and plan accordingly.

d. **Spatial Analysis:** Incorporate spatial analysis techniques to understand how population distribution within the watershed area may change over time. This includes mapping population density, identifying areas of rapid urbanization or depopulation, and assessing the spatial distribution of demographic characteristics.

e. **Socioeconomic Factors:** Consider socioeconomic factors that may influence future population trends, such as economic growth, employment opportunities, education levels, healthcare access, and environmental conditions. These factors can affect fertility rates, migration patterns, and overall population dynamics within watershed communities.

f. **Community Engagement:** Involve local stakeholders, including community members, government agencies, and non-governmental organizations, in the population trend analysis and projection process. Their input can provide valuable insights into local context, priorities, and concerns, helping to ensure that planning and development strategies are responsive to community needs.

By assessing past trends and projecting future population growth rates within watershed communities, decision-makers can better anticipate and plan for future challenges and opportunities related to infrastructure development, resource management, environmental conservation, and social services provision. These projections serve as valuable inputs for long-term planning efforts aimed at promoting sustainable and inclusive development within watershed areas.

4.2 Livelihood Patterns and Economic Activities

Primary Economic Sectors: Identifying the primary sources of livelihoods and economic activities within watershed communities, including agriculture, forestry, fishing, mining, tourism, and small-scale industries. Identifying the primary economic sectors within watershed communities is crucial for understanding the local economy, livelihoods, and resource utilization. Here are the key sectors commonly found within watershed areas:

a. **Agriculture:** Agriculture is often a primary economic activity within watershed communities. It includes crop cultivation, livestock rearing, and agroforestry practices. Farmers may grow a variety of crops suited to the local climate and soil conditions, including staples such as grains, fruits, vegetables, and cash crops like coffee, tea, or cocoa.

b. **Forestry:** Forestry plays a significant role in many watershed areas, where forests provide valuable resources such as timber, non-timber forest products (NTFPs), and ecosystem services like watershed protection and biodiversity conservation. Communities may engage in sustainable forestry practices, such as selective logging or forest management for ecotourism.

c. **Fishing:** Watershed communities located near rivers, lakes, or coastal areas often rely on fishing as a primary source of livelihood. Fishermen may engage in small-scale or artisanal fishing practices, using traditional or modern fishing techniques to catch fish and other aquatic resources for local consumption or sale.

d. **Mining:** Mining activities, including artisanal and small-scale mining (ASM), can be found in some watershed areas rich in mineral resources such as gold, diamonds, or gemstones. However, mining can also pose environmental and social challenges, including water pollution, land degradation, and conflicts over resource rights.

e. **Tourism:** Tourism can be a significant economic driver in watershed communities blessed with natural attractions such as mountains, rivers, forests, or cultural heritage sites. Tourism activities may include ecotourism, adventure tourism, cultural tourism, and recreational activities like hiking, wildlife viewing, or birdwatching.

f. **Small-Scale Industries:** Small-scale industries, such as handicrafts production, food processing, and manufacturing of locally made goods, provide employment and income opportunities for residents in many watershed communities. These industries often rely on locally available resources and traditional craftsmanship.

Identifying and understanding the primary economic sectors within watershed communities is essential for sustainable development planning, natural resource management, and poverty alleviation efforts. It enables policymakers, local authorities, and development practitioners to tailor interventions and investments that support inclusive economic growth, environmental conservation, and resilience-building within watershed areas.

Table: Table provides an overview of the various economic activities and livelihood patterns commonly found near watersheds.

S. No.	Agriculture	Cultivation of crops and livestock	- Small-scale farming	- Rice cultivation
			Subsistence agriculture	- Vegetable farming
			- Livestock rearing	- Dairy farming
	Fisheries	Fishing and aquaculture activities	- Small-scale fishery	- Inland fish farming
			Commercial fishing	- Fisheries cooperatives
			- Fish processing and trade	- Fish markets
	Forestry	Timber extraction and non-timber forest products	- Sustainable forestry	- Timber logging
			- Collection of forest products	- Bamboo harvesting
			- Agroforestry	- Honey production
	Tourism	Recreation and ecotourism services	- Nature-based tourism	- Wildlife safaris
			- Adventure tourism	- Trekking and hiking
			- Eco-lodges and accommodations	- Birdwatching tours
	Industry	Manufacturing and processing industries	- Small-scale cottage industries	- Handicraft production
			Food processing	- Rice mills
			- Agro-based industries	- Sugar mills
	Services	Provision of services such as transportation and retail	- Retail shops and markets	- Grocery stores
			- Transportation services	- Taxi services
			- Healthcare and education	- Clinics and schools
			- Hospitality and catering	- Restaurants and hotels

- **Livelihood Strategies:** Examining the diversity of livelihood strategies adopted by households, including subsistence farming, wage labor, entrepreneurship, and non-farm activities. Examining the diversity of livelihood strategies adopted by households within watershed communities provides insights into the economic resilience, vulnerability, and adaptation capacity of local populations. Here are the key livelihood strategies commonly observed:

  a. **Subsistence Farming:** Subsistence farming involves cultivating crops and raising livestock primarily for household consumption rather than for sale in markets. Households practicing subsistence farming often grow a variety of crops suited to local conditions and may engage in mixed farming or agroforestry to diversify their food sources and income streams.

  b. **Wage Labor:** Many households within watershed communities rely on wage labor for income generation. This can include formal employment in sectors such as agriculture, construction, manufacturing, or services, as well as informal or seasonal work opportunities. Wage labor provides a steady income source for households and may supplement other livelihood activities.

  c. **Entrepreneurship**: Entrepreneurship involves starting and running small businesses or enterprises to generate income and create employment opportunities. In watershed communities, entrepreneurship may take various forms, such as small-scale farming enterprises, retail shops, food stalls, artisanal crafts production, or eco-tourism ventures. Entrepreneurial activities contribute to economic diversification and local development.

  d. **Non-Farm Activities**: Non-farm activities encompass a wide range of income-generating activities outside of agriculture and wage labor. These may include small-scale trading, handicrafts production, transportation services, food processing, hospitality services, and other informal or micro-enterprises. Non-farm activities provide alternative livelihood options for households and contribute to rural economic diversification.

  e. **Natural Resource-based Activities**: Some households within watershed communities rely on natural resource-based activities such as fishing, hunting, gathering, or collection of non-timber forest products (NTFPs) for their livelihoods. These activities are often closely linked to local ecosystems and cultural traditions, and they play a vital role in supporting rural livelihoods and preserving traditional knowledge.

f. **Remittances**: Remittances from family members working outside the watershed area or abroad can also contribute significantly to household incomes in some communities. Remittances are often used to support education, healthcare, housing, and other household expenses, providing a lifeline for families in rural areas.

Examining the diversity of livelihood strategies adopted by households within watershed communities requires a holistic understanding of local socio-economic dynamics, natural resource availability, market opportunities, and external factors such as climate change and policy environments. Supporting sustainable livelihoods requires targeted interventions that enhance access to productive resources, build skills and capacities, promote entrepreneurship, and strengthen social safety nets within watershed areas.

- **Income and Poverty Levels:** Assessing income distribution patterns and poverty levels within watershed communities, including disparities between different socio-economic groups and marginalized populations. Assessing income distribution patterns and poverty levels within watershed communities is essential for understanding socio-economic disparities, identifying vulnerable populations, and designing targeted interventions to alleviate poverty and promote inclusive development. Here are the key steps involved in this assessment:

a. **Income Distribution Analysis**: Begin by collecting data on household incomes within the watershed area. This may involve conducting household surveys, analyzing census data, or using secondary sources such as national statistics. Analyze the distribution of income levels to identify patterns of inequality and disparity within the community.

b. **Poverty Assessment**: Use poverty indicators such as income poverty lines, multidimensional poverty indices, or poverty headcount ratios to assess the prevalence and severity of poverty within watershed communities. This involves identifying households living below the poverty line and understanding the factors contributing to their poverty status.

c. **Socio-economic Disparities**: Examine socio-economic disparities within watershed communities by analyzing income levels, asset ownership, access to basic services (such as education, healthcare, and sanitation), and employment opportunities. Identify marginalized populations, including ethnic minorities, indigenous communities, women, children, the elderly, persons with disabilities, and other vulnerable groups facing barriers to social and economic inclusion.

d. **Spatial Analysis:** Conduct spatial analysis to identify geographic patterns of poverty and socio-economic disparities within the watershed area. Mapping poverty hotspots, disadvantaged areas, and marginalized communities helps prioritize interventions and target resources where they are most needed.

e. **Intersectional Analysis:** Consider intersecting factors such as gender, ethnicity, age, and disability status in the assessment of income and poverty levels. Intersectional analysis helps identify how multiple forms of discrimination and marginalization intersect to exacerbate poverty and exclusion for certain groups within the community.

f. **Participatory Approaches:** Engage with local communities, stakeholders, and marginalized groups in participatory approaches to poverty assessment. Participatory methods such as community meetings, focus group discussions, and participatory wealth ranking help ensure that the voices and perspectives of diverse community members are heard and integrated into poverty alleviation strategies.

g. **Monitoring and Evaluation:** Establish mechanisms for monitoring and evaluating poverty alleviation efforts within watershed communities. Regularly track progress against poverty reduction targets, assess the impact of interventions on income levels and well-being, and adjust strategies as needed based on feedback from stakeholders and beneficiaries.

By assessing income distribution patterns and poverty levels within watershed communities, policymakers, development practitioners, and local authorities can design evidence-based interventions that address root causes of poverty, promote economic empowerment, and enhance social inclusion for all members of the community.

- **Value Chains and Market Access:** Analyzing the functioning of value chains for key commodities and the access of watershed communities to markets, credit, technology, and infrastructure. Analyzing the functioning of value chains for key commodities and the access of watershed communities to markets, credit, technology, and infrastructure is essential for promoting economic development, enhancing livelihoods, and reducing poverty. Here's how this analysis can be conducted:

a. **Value Chain Analysis:** Start by identifying key commodities or products produced within watershed communities, such as agricultural crops, forestry products, fishery products, handicrafts, or tourism services. Conduct a value chain analysis to understand the entire process from production to consumption, including inputs, production processes, value addition, distribution channels, and market outlets.

b. **Market Assessment:** Assess market dynamics and demand for products originating from watershed communities. Identify market trends, consumer preferences, price fluctuations, and market opportunities for local producers. Analyze domestic and international markets, as well as regional trade networks, to identify potential market niches and export opportunities.

c. **Access to Markets:** Evaluate the accessibility of markets for watershed communities, considering factors such as transportation infrastructure, road networks, market facilities, and distance to major urban centers or export hubs. Identify barriers to market access, including poor road conditions, high transportation costs, limited market information, and lack of market linkages.

d. **Access to Credit:** Examine access to financial services, including credit, savings, and insurance, for producers and entrepreneurs within watershed communities. Assess the availability of formal financial institutions, microfinance institutions, community-based savings groups, and other sources of credit. Identify barriers to access, such as collateral requirements, interest rates, and financial literacy.

e. **Access to Technology:** Analyze access to technology and innovation within watershed communities, including agricultural technologies, processing equipment, information and communication technologies (ICTs), and renewable energy solutions. Evaluate the adoption of modern farming practices, mechanization, value-added processing techniques, and digital platforms for market information and e-commerce.

f. **Access to Infrastructure:** Assess the availability and quality of infrastructure necessary for economic activities within watershed communities, including transportation infrastructure (roads, bridges, ports), energy infrastructure (electricity, renewable energy), water supply and sanitation facilities, and market infrastructure (wholesale markets, storage facilities, cold chains).

g. **Stakeholder Engagement:** Engage with key stakeholders, including local producers, traders, cooperatives, government agencies, financial institutions, and development organizations, in the analysis of value chains and market access. Collaboration with stakeholders helps identify opportunities for collaboration, leverage resources, and address challenges collectively.

By analyzing the functioning of value chains and assessing access to markets, credit, technology, and infrastructure within watershed communities, policymakers and development practitioners can design targeted interventions to strengthen local economies, enhance market integration, and improve livelihood opportunities for community members. This analysis forms the basis for inclusive and sustainable economic development strategies tailored to the specific needs and context of watershed areas.

4.3 Social Institutions and Governance Structures

- **Community Organizations:** Identifying formal and informal community-based organizations within watershed areas, including self-help groups, cooperatives, user groups, and village councils. Identifying formal and informal community-based organizations within watershed areas is crucial for understanding local governance structures, social networks, and community resilience. Here are the key types of community organizations commonly found within watershed communities:
  a. **Self-Help Groups (SHGs):** Self-help groups are informal associations of individuals, usually women, who come together to address common social, economic, or livelihood challenges. SHGs often focus on savings and credit activities, skill development, income-generating activities, and social support networks. They play a vital role in empowering women, enhancing financial inclusion, and promoting community solidarity.
  b. **Cooperatives:** Cooperatives are formal organizations owned and operated by their members to meet common economic, social, and cultural needs. In watershed areas, cooperatives may include agricultural cooperatives, producer cooperatives, marketing cooperatives, savings and credit cooperatives (SACCOs), and forestry cooperatives. Cooperatives enable small-scale producers to access markets, credit, inputs, and services collectively, enhancing their bargaining power and economic resilience.
  c. **User Groups:** User groups are community-based organizations responsible for managing and governing shared natural resources such as water, forests, grazing lands, and fisheries within watershed areas. These groups may include water user associations (WUAs), forest user groups (FUGs), community fisheries management

committees, and pasture user committees. User groups play a crucial role in sustainable natural resource management, conflict resolution, and ecosystem conservation.

d. **Village Councils or Community Committees:** Village councils, community committees, or community development associations are local governance structures representing residents within watershed communities. These entities may be formalized through local government regulations or operate informally based on traditional or customary practices. Village councils often play a role in decision-making, planning, and implementation of community development initiatives, including infrastructure projects, social programs, and natural resource management activities.

e. **Women's Groups and Youth Associations:** Women's groups and youth associations are organized platforms for women and young people to mobilize, advocate for their rights, and address gender-specific or youth-related issues within watershed communities. These groups may engage in income-generating activities, skill development programs, health awareness campaigns, and community development projects tailored to the needs of women and youth.

Identifying formal and informal community-based organizations within watershed areas requires engaging with local stakeholders, conducting community assessments, and mapping existing organizations and their roles. Understanding the functions, strengths, and challenges of these organizations is essential for promoting participatory decision-making, enhancing social capital, and fostering sustainable development within watershed communities. Collaboration with community organizations enables stakeholders to leverage local knowledge, resources, and capacities in addressing common challenges and achieving shared goals.

Table: Table provides an overview of the diverse social institutions and governance structures involved in watershed management in India.

S. No.	Social Institution / Governance Structure	Description	Roles and Responsibilities	Examples
1	Watershed Committees	Local community-based organizations	- Participating in decision-making processes	- Village Watershed Committees
			- Implementing watershed management activities	- Gram Panchayat Committees
			- Mobilizing resources and labor for projects	- Self-Help Groups (SHGs)
2	User Groups	Groups representing specific user interests	- Representing the interests of specific stakeholders	- Farmers' Associations
			- Collaborating with government agencies	- Fishermen Cooperatives
			- Implementing community-based projects	- Women's Self-Help Groups (SHGs)
3	Watershed Development Agencies	Government agencies responsible for watershed projects	- Planning and implementing watershed development	- Watershed Development Departments
			- Providing technical support and funding	- Watershed Development Corporations (WDCs)
			- Monitoring and evaluating project outcomes	- State Departments of Rural Development
4	Research Institutions	Academic and research organizations	- Conducting studies on watershed dynamics	- Universities
			- Providing technical expertise and training	- Research Institutes
			- Collaborating on research projects	- National Environmental Research Institutes (NERIs)
5	NGOs and Civil Society Organizations	Non-profit organizations working on social and environmental issues	- Advocating for sustainable watershed management	- Environmental NGOs

			- Raising awareness and mobilizing communities	- Community-Based Organizations (CBOs)
			- Implementing community-led projects	- Watershed NGOs
6	Government Departments	Various government agencies involved in watershed management	- Formulating policies and regulations	- Ministry of Environment, Forests and Climate Change
			- Implementing watershed development programs	- Departments of Agriculture, Horticulture, and Animal Husbandry
			- Monitoring and enforcing regulatory measures	- Departments of Water Resources
			- Providing funding and technical support	- State Pollution Control Boards
7	Financial Institutions	Banks and financial organizations	- Providing loans and financial assistance	- Regional Rural Banks
			- Investing in watershed development projects	- Microfinance Institutions (MFIs)
			- Supporting income-generating activities	- Cooperative Banks

- **Traditional Institutions:** Exploring the role of traditional governance structures, customary laws, and indigenous knowledge systems in managing natural resources and resolving conflicts within watershed communities.Exploring the role of traditional governance structures, customary laws, and indigenous knowledge systems in managing natural resources and resolving conflicts within watershed communities provides valuable insights into local governance dynamics, social cohesion, and sustainable resource management. Here's how traditional institutions contribute to community resilience within watershed areas:
  a. **Traditional Governance Structures:** Traditional governance structures, such as chiefs, elders, councils of elders, or tribal leaders, play a significant role in decision-making, conflict resolution, and community leadership within watershed communities. These institutions often embody local customs, values, and norms and are respected for their authority and legitimacy. Traditional leaders may oversee resource management practices, mediate disputes, and uphold customary laws governing land use, water allocation, and natural resource extraction.
  b. **Customary Laws and Norms:** Customary laws and norms govern social behavior, land tenure, resource access, and community relationships within watershed areas. These unwritten rules and traditions are passed down through generations and reflect the collective wisdom and cultural heritage of local communities. Customary laws may regulate activities such as farming practices, grazing rights, fishing rights, and access to water sources, ensuring equitable distribution and sustainable use of natural resources.
  c. **Indigenous Knowledge Systems:** Indigenous knowledge systems encompass traditional ecological knowledge, local practices, and wisdom accumulated over generations by indigenous and local communities. This knowledge is often based on close observation of the natural environment and reflects an intimate understanding of local ecosystems, biodiversity, weather patterns, and ecological processes. Indigenous knowledge systems inform decision-making related to agriculture, forestry, water management, and disaster risk reduction, contributing to ecosystem resilience and adaptation to environmental changes.
  d. **Conflict Resolution Mechanisms:** Traditional institutions and customary laws play a crucial role in resolving conflicts and disputes within watershed communities. Traditional conflict resolution

mechanisms may involve mediation by community elders, customary courts, or consensus-based decision-making processes grounded in local traditions and cultural values. These mechanisms prioritize reconciliation, restoration of social harmony, and preservation of community cohesion, contributing to conflict prevention and peacebuilding within watershed areas.

e. **Partnership with Formal Institutions:** Traditional institutions often complement and coexist with formal governance structures, including local governments, regulatory agencies, and non-governmental organizations. Collaborative partnerships between traditional and formal institutions can leverage the strengths of both systems, enhance mutual understanding, and promote inclusive and participatory approaches to natural resource management and community development.

Exploring the role of traditional institutions in managing natural resources and resolving conflicts within watershed communities requires respectful engagement with local cultures, traditions, and knowledge systems. Integrating traditional governance structures and indigenous knowledge into broader governance frameworks can contribute to more holistic and contextually appropriate approaches to watershed management, fostering sustainability, resilience, and social justice within communities.

- **Local Governance Mechanisms:** Assessing the effectiveness of local government institutions, such as gram panchayats, watershed committees, and local development councils, in participatory decision-making and resource management. Assessing the effectiveness of local government institutions in participatory decision-making and resource management is crucial for promoting good governance, community empowerment, and sustainable development within watershed areas. Here's how the effectiveness of various local governance mechanisms, such as gram panchayats, watershed committees, and local development councils, can be assessed:

a. **Gram Panchayats:** Gram panchayats are local government bodies at the village level in India responsible for local administration and development. To assess their effectiveness in watershed management, consider factors such as:

- **Representation and inclusivity:** Evaluate the extent to which gram panchayats represent diverse social groups, including women, marginalized communities, and indigenous peoples.

- **Capacity and resources:** Assess the capacity of gram panchayats in terms of human resources, financial resources, and technical expertise to undertake watershed management activities effectively.
- **Decision-making processes**: Analyze the transparency, accountability, and participatory nature of decision-making processes within gram panchayats, including mechanisms for community engagement and consultation.
- **Implementation and monitoring:** Evaluate the effectiveness of gram panchayats in implementing watershed management projects, monitoring progress, and ensuring accountability for outcomes.

b. **Watershed Committees**: Watershed committees are community-based organizations responsible for coordinating and implementing watershed management activities at the local level. To assess their effectiveness, consider:

- **Composition and representation:** Examine the composition of watershed committees, including representation from diverse stakeholder groups such as farmers, landowners, women, and youth.
- **Functioning and decision-making**: Evaluate the functioning of watershed committees in terms of leadership, decision-making processes, and collaboration with other stakeholders, including government agencies and non-governmental organizations.
- **Capacity building and training:** Assess the capacity of watershed committees in terms of technical knowledge, skills development, and access to training and support services for watershed management.
- **Sustainability and resilience**: Analyze the long-term sustainability and resilience of watershed committees in mobilizing community resources, building social capital, and adapting to changing environmental conditions.

c. **Local Development Councils:** Local development councils or similar bodies may exist at higher administrative levels, such as block or district levels, to coordinate development activities across multiple villages or watersheds. To assess their effectiveness, consider:

- **Coordination and collaboration:** Evaluate the role of local development councils in coordinating watershed management efforts across multiple gram panchayats or administrative units, fostering collaboration among stakeholders, and leveraging resources effectively.

- **Planning and prioritization:** Assess the effectiveness of local development councils in facilitating participatory planning processes, setting priorities for watershed management interventions, and allocating resources based on local needs and priorities.
- **Monitoring and evaluation:** Analyze the mechanisms for monitoring and evaluating the performance of local development councils, including indicators of progress, feedback mechanisms, and mechanisms for accountability and transparency.

In assessing the effectiveness of local government institutions in participatory decision-making and resource management within watershed areas, it is essential to engage with local stakeholders, including community members, government officials, civil society organizations, and development partners. Participatory approaches, including community consultations, focus group discussions, and participatory appraisal methods, can help ensure that assessments are inclusive, contextually relevant, and sensitive to local perspectives and priorities.

Civil Society Engagement: Examining the involvement of non-governmental organizations (NGOs), civil society groups, and advocacy networks in promoting social mobilization, awareness raising, and capacity building within watershed communities.

Examining the involvement of non-governmental organizations (NGOs), civil society groups, and advocacy networks in promoting social mobilization, awareness raising, and capacity building within watershed communities in India reveals their significant contributions to sustainable development and natural resource management. Here's how civil society engages with watershed communities:

a. **Community Mobilization and Empowerment:** NGOs and civil society organizations play a vital role in mobilizing and empowering communities within watershed areas. They facilitate community meetings, awareness campaigns, and participatory decision-making processes to build consensus, foster social cohesion, and empower marginalized groups, including women, youth, and indigenous communities.

b. **Capacity Building and Training:** Civil society organizations provide capacity building and training programs to enhance the knowledge, skills, and capabilities of local stakeholders in watershed management practices. These programs cover a range of topics, including sustainable agriculture, water conservation, soil conservation, biodiversity conservation, climate change adaptation, and disaster risk reduction.

c. **Technical Support and Expertise:** NGOs and civil society groups offer technical support and expertise to watershed communities in designing and implementing watershed management projects. They provide technical assistance in areas such as land use planning, watershed mapping, soil and water conservation techniques, afforestation, sustainable agriculture practices, and water harvesting technologies.

d. **Advocacy and Policy Influence:** Civil society organizations engage in advocacy and policy dialogue to promote favorable policies, laws, and regulations related to watershed management, environmental conservation, and rural development. They advocate for the rights and interests of local communities, raise awareness about environmental issues, and mobilize public support for sustainable development initiatives.

e. **Partnerships and Networking:** NGOs, civil society groups, and advocacy networks often collaborate with government agencies, research institutions, academic organizations, and international development partners to leverage resources, share best practices, and scale up successful interventions. They participate in multi-stakeholder platforms, knowledge exchange forums, and collaborative initiatives aimed at promoting integrated watershed management and sustainable livelihoods.

f. **Monitoring and Evaluation:** Civil society organizations contribute to monitoring and evaluation efforts within watershed communities by collecting data, conducting assessments, and documenting best practices and lessons learned. They monitor the impact of watershed management projects on environmental sustainability, livelihoods, social equity, and community resilience, and use this information to inform decision-making and adaptive management processes.

Examining civil society engagement in promoting social mobilization, awareness raising, and capacity building within watershed communities in India requires a comprehensive understanding of the diverse roles, approaches, and contributions of NGOs, civil society groups, and advocacy networks. By working in partnership with government agencies, local communities, and other stakeholders, civil society organizations play a crucial role in fostering participatory governance, empowering communities, and advancing sustainable development goals within watershed areas.

4.4 Vulnerability and Resilience of Communities to Environmental Changes

Environmental Risks and Hazards: Identifying environmental risks and hazards faced by watershed communities, including floods, droughts, landslides, water pollution, deforestation, and climate variability. Identifying environmental risks and hazards faced by watershed communities is crucial for developing effective risk reduction and management strategies. Here are some key environmental risks and hazards commonly encountered in watershed areas:

a. **Floods:** Watershed communities are often vulnerable to flooding, especially during the monsoon season when heavy rainfall can lead to river overflow, flash floods, and inundation of low-lying areas. Floods can cause damage to homes, infrastructure, agricultural fields, and disrupt livelihood activities, posing risks to lives and livelihoods.

b. **Droughts:** Droughts are another significant environmental risk for watershed communities, particularly in regions with irregular rainfall patterns or prolonged dry spells. Droughts can lead to water scarcity, crop failures, livestock deaths, and food insecurity, exacerbating poverty and livelihood vulnerabilities.

c. **Landslides:** Steep slopes and unstable terrain in watershed areas make communities susceptible to landslides, especially during periods of heavy rainfall or seismic activity. Landslides can cause loss of life, damage to infrastructure, destruction of agricultural land, and disruption of transportation routes, posing risks to human settlements and economic activities.

d. **Water Pollution**: Pollution of water sources, including rivers, lakes, and groundwater, poses significant health and environmental risks for watershed communities. Pollution from industrial effluents, agricultural runoff, solid waste disposal, and untreated sewage can contaminate water supplies, degrade ecosystems, and threaten public health through waterborne diseases.

e. **Deforestation:** Deforestation and forest degradation within watersheds contribute to environmental risks such as soil erosion, loss of biodiversity, reduced water retention capacity, and increased vulnerability to natural disasters. Deforestation also undermines ecosystem services such as watershed protection, carbon sequestration, and climate regulation, affecting the resilience of communities to environmental shocks.

f. **Climate Variability**: Climate variability, including changes in temperature, precipitation patterns, and extreme weather events, poses risks to watershed communities by altering local ecosystems, water availability, agricultural productivity, and natural resource-dependent livelihoods. Climate change exacerbates existing vulnerabilities and may lead to more frequent and intense environmental hazards in the future.

Identifying environmental risks and hazards within watershed communities requires conducting vulnerability assessments, hazard mapping, and community consultations to understand local context and priorities. Once identified, stakeholders can develop risk reduction and management plans that integrate measures such as early warning systems, disaster preparedness training, land use planning, reforestation efforts, water conservation practices, and climate-resilient livelihood strategies. Collaboration among government agencies, civil society organizations, research institutions, and local communities is essential for implementing effective solutions to mitigate environmental risks and build community resilience within watershed areas.

- **Social Vulnerability Factors:** Assessing socio-economic factors that contribute to the vulnerability of communities, such as poverty, inequality, lack of access to basic services, and marginalization. Assessing social vulnerability factors is essential for understanding the underlying drivers of vulnerability within watershed communities and developing targeted interventions to address them. Here are key socio-economic factors that contribute to community vulnerability:

  a. **Poverty:** Poverty is a significant social vulnerability factor that undermines the resilience of communities within watershed areas. Poverty limits access to resources, services, and opportunities, exacerbating the impacts of environmental hazards and hindering adaptive capacity. Poor households are often more reliant on natural resources for their livelihoods, making them particularly vulnerable to environmental shocks and disruptions.

  b. **Inequality:** Social and economic inequality exacerbates vulnerability within watershed communities by disproportionately affecting marginalized groups, such as women, ethnic minorities, indigenous peoples, and persons with disabilities. Unequal access to resources, opportunities, and decision-making power can amplify the impacts of environmental hazards and limit the ability of marginalized populations to cope and recover from disasters.

c. **Lack of Access to Basic Services**: Limited access to essential services such as clean water, sanitation, healthcare, education, and social protection increases vulnerability to environmental risks and undermines community resilience. Communities without access to basic services face greater health risks, reduced economic opportunities, and lower levels of human development, making them more susceptible to the impacts of environmental hazards.

d. **Marginalization:** Marginalized communities, including rural populations, indigenous peoples, and those living in remote or ecologically fragile areas, are often more vulnerable to environmental risks and hazards due to factors such as geographic isolation, limited infrastructure, and historical exclusion from decision-making processes. Marginalized groups may face barriers to accessing resources, information, and support services, further increasing their vulnerability to environmental stressors.

e. **Livelihood Dependence on Natural Resources**: Communities that rely heavily on natural resources for their livelihoods, such as agriculture, forestry, fishing, and pastoralism, are particularly vulnerable to environmental changes and disruptions. Dependence on rain-fed agriculture, for example, makes communities susceptible to droughts and erratic rainfall patterns, while reliance on forest resources exposes them to risks such as deforestation, land degradation, and loss of biodiversity.

f. **Lack of Social Capital**: Social cohesion, community networks, and collective action are important sources of resilience within watershed communities. Lack of social capital, including trust, reciprocity, and cooperation, can weaken community resilience and hinder effective response and recovery efforts in the face of environmental hazards.

Assessing social vulnerability factors within watershed communities requires a multidimensional approach that considers the interplay of socio-economic, cultural, and institutional factors. By understanding the root causes of vulnerability, stakeholders can develop targeted interventions that address underlying inequalities, strengthen social protection mechanisms, build adaptive capacity, and empower communities to cope with and adapt to environmental risks and hazards.

Adaptive Capacities: Examining the adaptive capacities of communities to cope with and respond to environmental changes, including traditional knowledge, social networks, livelihood diversification, and access to information and resources. Examining the adaptive capacities of communities to cope with and respond to environmental changes is crucial for understanding their resilience and for designing effective adaptation strategies. Here are key aspects to consider when assessing adaptive capacities within watershed communities:

a. **Traditional Knowledge**: Traditional ecological knowledge (TEK) is accumulated wisdom passed down through generations within communities. It encompasses insights, practices, and beliefs about local ecosystems, natural resources, weather patterns, and sustainable land management. Examining the extent to which communities draw on traditional knowledge for decision-making and adaptive responses provides valuable insights into their resilience.

b. **Social Networks:** Social networks play a vital role in facilitating information exchange, resource sharing, and collective action within communities. Strong social networks enable communities to mobilize resources, support vulnerable members, and coordinate responses to environmental changes and disasters. Assessing the structure, strength, and inclusivity of social networks helps identify sources of social capital and resilience within watershed communities.

c. **Livelihood Diversification:** Communities with diversified livelihood strategies are often more resilient to environmental shocks and stresses. Livelihood diversification involves engaging in multiple income-generating activities, such as agriculture, livestock rearing, non-farm enterprises, and off-farm employment. Examining the extent to which communities diversify their livelihoods and adapt to changing environmental conditions provides insights into their adaptive capacities.

d. **Access to Information and Resources:** Access to timely and relevant information, as well as resources such as financial capital, technology, and infrastructure, is essential for adaptive capacity building. Communities with access to weather forecasts, early warning systems, extension services, and market information can make informed decisions and take proactive measures to mitigate risks and capitalize on opportunities. Assessing barriers to access and opportunities for improving information and resource availability helps identify pathways for enhancing adaptive capacities.

e. **Institutional Support**: Supportive institutional frameworks and governance structures facilitate adaptive responses to environmental changes. Local governments, community-based organizations, NGOs, and other stakeholders play a role in providing technical assistance, financial support, and policy guidance to communities. Examining the effectiveness of institutional support mechanisms and their alignment with community needs and priorities helps strengthen adaptive capacities.

f. **Innovative Practices:** Communities often develop innovative practices and technologies to cope with environmental challenges and exploit new opportunities. These innovations may include sustainable agriculture techniques, water harvesting systems, agroforestry models, and community-based disaster risk reduction initiatives. Identifying and documenting innovative practices within watershed communities informs learning, replication, and scaling-up efforts.

Examining the adaptive capacities of communities within watershed areas requires a participatory and context-specific approach that engages community members, stakeholders, and experts. By understanding the strengths, vulnerabilities, and potential for adaptation within communities, stakeholders can develop targeted interventions that enhance resilience and promote sustainable development in the face of environmental changes.

Building Resilience: Discussing strategies for building resilience and enhancing the adaptive capacities of watershed communities through targeted interventions, capacity building, risk reduction measures, and social protection mechanisms. Building resilience and enhancing the adaptive capacities of watershed communities requires a holistic approach that integrates various strategies and interventions tailored to local contexts and priorities. Here are key strategies for building resilience within watershed communities:

a. **Community-Based Disaster Risk Reduction (CBDRR):** Implement community-based disaster risk reduction initiatives that empower local communities to identify, assess, and mitigate risks associated with environmental hazards such as floods, droughts, landslides, and water pollution. CBDRR strategies may include early warning systems, contingency planning, hazard mapping, infrastructure upgrades, and natural resource management practices that enhance ecosystem resilience.

b. **Capacity Building and Skill Development**: Provide capacity building and skill development programs to enhance the knowledge, skills, and capabilities of community members in watershed management, sustainable agriculture, climate-smart practices, disaster preparedness, and livelihood diversification. Training workshops, educational programs, experiential learning activities, and peer-to-peer exchanges can strengthen adaptive capacities and promote innovation within communities.

Case Study 1: Capacity Building and Skill Development in the Indian Himalayas

Background

In the Indian Himalayas, communities are heavily dependent on agriculture, forestry, and tourism for their livelihoods. These areas are particularly vulnerable to climate change, land degradation, and natural disasters. The Watershed Organisation Trust (WOTR) implemented a comprehensive capacity building and skill development program to enhance the adaptive capacities of these communities, focusing on sustainable watershed management and climate-resilient practices.

Project Implementation

Location: Uttarakhand and Himachal Pradesh, Indian Himalayas

Key Components

Training Workshops

- Conducted on topics such as sustainable agriculture, climate-smart practices, water conservation, and disaster preparedness.
- Experts from agricultural universities and research institutions provided hands-on training.

Educational Programs:

- Developed curriculums for schools and adult education centers focused on environmental sustainability and watershed management.
- Integrated climate change awareness and adaptation strategies into existing educational frameworks.

Experiential Learning Activities

- Organized field visits to model farms practicing sustainable agriculture and water management.
- Implemented demonstration projects on rainwater harvesting, contour trenching, and agroforestry.

Peer-to-Peer Exchanges

- Facilitated knowledge exchange visits between farmers from different regions to share best practices and innovations.
- Created local farmer groups and networks to sustain peer learning and support.
- Outcomes and Impacts

Enhanced Knowledge and Skills

- Over 5,000 community members trained in sustainable agricultural practices, leading to increased crop yields and diversified farming systems.
- Improved understanding of water conservation techniques, resulting in more efficient use of water resources.

Increased Economic Resilience

- Diversification of income sources through the adoption of agroforestry, bee-keeping, and eco-tourism activities.
- Establishment of small-scale enterprises, such as processing units for medicinal plants and local crafts, creating new employment opportunities.

Improved Adaptive Capacity

- Development of community-based disaster preparedness plans and early warning systems.
- Strengthened local institutions and governance structures to support sustainable watershed management.

Empowerment of Women and Marginalized Groups

- Special focus on empowering women through targeted training programs in microfinance and entrepreneurship.
- Formation of women's self-help groups to manage savings and credit activities, facilitating financial inclusion and independence.
- Lessons Learned

Community Participation

- Active involvement of community members in planning and implementation processes ensured the relevance and acceptance of the programs.
- Building trust and fostering collaboration among stakeholders were crucial for the success of the initiatives.

Integration of Traditional Knowledge

- Combining scientific knowledge with traditional practices enhanced the effectiveness of the interventions.
- Recognizing and valuing local knowledge fostered greater community ownership and sustainability of the projects.

Continuous Monitoring and Adaptation

- Regular monitoring and evaluation of the programs allowed for adaptive management and refinement of strategies.
- Flexibility to adapt to changing conditions and feedback from the community ensured the long-term impact of the capacity building efforts.

Conclusion

The capacity building and skill development programs implemented by WOTR in the Indian Himalayas demonstrate the transformative potential of empowering communities through knowledge and skills enhancement. By focusing on sustainable practices and livelihood diversification, these initiatives have significantly improved the socio-economic conditions and resilience of watershed communities, providing a model for similar efforts in other regions.

c. **Natural Resource Management**: Promote sustainable natural resource management practices that conserve and restore ecosystems, enhance biodiversity, and improve ecosystem services within watershed areas. Strategies may include afforestation and reforestation efforts, soil and water conservation measures, sustainable land use planning, watershed restoration projects, and integrated water resources management approaches that balance competing demands for water.

d. **Social Protection Mechanisms**: Establish social protection mechanisms that provide safety nets for vulnerable populations within watershed communities, including social assistance programs, cash transfers, food security initiatives, health insurance schemes, and asset-building programs. Social protection mechanisms help mitigate the impacts of environmental shocks and stresses, reduce poverty and inequality, and enhance resilience to future risks.

e. **Livelihood Diversification and Income Generation:** Support livelihood diversification and income generation activities that reduce dependency on single sources of income and increase economic resilience within watershed communities. This may involve promoting alternative livelihood options such as agroforestry, eco-tourism, small-scale enterprises, value-added processing, and microfinance initiatives that empower women and marginalized groups.

f. **Participatory Governance and Decision-Making**: Foster participatory governance processes that engage local communities, stakeholders, and marginalized groups in decision-making, planning, and implementation of development initiatives within watershed areas. Participatory approaches promote ownership, accountability, and inclusivity, ensuring that interventions are contextually relevant, socially equitable, and sustainable over the long term.

g. **Innovative Technologies and Practices**: Harness innovative technologies and practices that leverage local knowledge, traditional wisdom, and modern advancements to address environmental challenges and enhance resilience within watershed communities. This may include the adoption of climate-resilient crops, water-saving technologies, renewable energy solutions, digital tools for information dissemination, and nature-based solutions that capitalize on ecosystem services.

h. **Collaborative Partnerships and Networks**: Foster collaborative partnerships and networks among government agencies, civil society organizations, research institutions, private sector actors, and community members to leverage resources, share expertise, and scale up successful interventions for building resilience within watershed communities. Multi-stakeholder platforms, knowledge exchange forums, and joint action plans facilitate collective action and learning.

By implementing these strategies for building resilience and enhancing adaptive capacities within watershed communities, stakeholders can promote sustainable development, reduce vulnerability to environmental risks, and improve the well-being and livelihoods of community members now and in the future.

Important Questions

1. What specific questions do you have regarding the paragraph on civil society engagement in promoting social mobilization, awareness raising, and capacity building within watershed communities in India?
2. How do NGOs and civil society organizations mobilize local communities within watershed areas?
3. Can you provide examples of capacity building programs implemented by civil society groups in Indian watershed communities?
4. What types of technical support do NGOs and civil society groups offer to watershed communities in India?
5. How do civil society organizations advocate for policy changes related to watershed management and environmental conservation?
6. Could you explain how partnerships and networking contribute to the effectiveness of civil society interventions in watershed communities?

7. What methods do civil society organizations use to monitor and evaluate the impact of their projects in Indian watershed communities?
8. How do civil society groups ensure the participation of marginalized groups, such as women and indigenous communities, in watershed management initiatives?
9. Can you describe a successful example of civil society engagement in promoting sustainable livelihoods in an Indian watershed community?
10. What are some of the challenges faced by civil society organizations in working with government agencies and other stakeholders in watershed management?
11. How do civil society organizations leverage local knowledge and cultural practices in their interventions within Indian watershed communities?
12. What specific question do you have regarding the strategies for building resilience and enhancing adaptive capacities within watershed communities outlined in the paragraphs above?
13. How do community-based disaster risk reduction initiatives empower local communities to identify and mitigate risks specific to their watershed areas?
14. Can you provide examples of capacity building programs that have successfully enhanced the resilience of watershed communities?
15. What are some effective natural resource management practices that contribute to ecosystem resilience and sustainable development within watershed areas?
16. How do social protection mechanisms contribute to reducing vulnerability and enhancing resilience among vulnerable populations within watershed communities?
17. What are the key challenges and opportunities associated with promoting livelihood diversification and income generation activities within watershed communities?
18. How can participatory governance processes ensure that development interventions are inclusive, equitable, and responsive to the needs of watershed communities?
19. What role can innovative technologies and practices play in addressing environmental challenges and enhancing resilience within watershed areas?
20. How do collaborative partnerships and networks facilitate the scaling up of successful resilience-building interventions and promote collective action within watershed communities?

21. Can you discuss the importance of monitoring and evaluation in assessing the effectiveness of resilience-building strategies and informing adaptive management approaches within watershed areas?
22. What are some examples of successful resilience-building initiatives implemented in watershed communities, and what lessons can be learned from their experiences?
23. What strategies can be employed to effectively promote alternative livelihood options such as agroforestry, eco-tourism, and small-scale enterprises within watershed communities, and how can these strategies enhance economic resilience?
24. How can microfinance initiatives and value-added processing be leveraged to empower women and marginalized groups in watershed communities, and what are the potential socio-economic impacts of these initiatives?
25. What methods can be used to assess the socio-economic conditions of people living in watershed communities, and how can this information inform the development of targeted livelihood diversification and income generation programs?

5

Integrated Watershed Management for Sustainable Development

5.1 Participatory Approaches to Watershed Management

Community Engagement Strategies: Exploring methods for involving local communities in decision-making processes, project planning, and implementation, including participatory rural appraisal (PRA), stakeholder consultations, and community meetings. Exploring methods for involving local communities in decision-making processes, project planning, and implementation is essential for promoting ownership, sustainability, and effectiveness of development initiatives within watershed areas. Here are key community engagement strategies:

a. **Participatory Rural Appraisal (PRA):** PRA is a participatory approach that involves community members in assessing their own needs, priorities, resources, and capacities. PRA techniques, such as transect walks, social mapping, seasonal calendars, and wealth ranking, enable communities to identify and analyze local issues, strengths, and opportunities. PRA fosters dialogue, builds trust, and empowers communities to take ownership of development processes.

b. **Stakeholder Consultations:** Conduct stakeholder consultations to engage a diverse range of actors, including community members, local leaders, government officials, civil society organizations, and other stakeholders. Consultations provide opportunities for stakeholders to share their perspectives, knowledge, and concerns, and contribute to decision-making and project design. Stakeholder consultations ensure inclusivity, transparency, and legitimacy in development processes.

c. **Community Meetings:** Organize community meetings, village assemblies, or public forums to facilitate dialogue, consensus-building, and decision-making among community members. Community meetings provide platforms for discussing development priorities, reviewing project progress, and resolving conflicts collaboratively. They foster social cohesion, collective action, and community mobilization around shared goals and aspirations.

d. **Focus Group Discussions (FGDs)**: Conduct focus group discussions with specific segments of the community, such as women, youth, farmers, or marginalized groups, to gather in-depth insights into their needs, perspectives, and priorities. FGDs provide a safe and supportive environment for participants to express their views, share experiences, and propose solutions to address their concerns. FGDs help ensure that diverse voices are heard and integrated into decision-making processes.

e. **Community-Based Organizations (CBOs):** Collaborate with local community-based organizations, such as self-help groups, cooperatives, user groups, and village councils, to facilitate community engagement and participation. CBOs serve as intermediaries between communities and external stakeholders, mobilize resources, and provide platforms for community empowerment and representation.

f. **Capacity Building Workshops**: Organize capacity building workshops, training sessions, and skill development programs to enhance the knowledge, skills, and capabilities of community members in areas such as project management, leadership, communication, and advocacy. Capacity building workshops empower communities to actively participate in decision-making, project planning, and implementation processes.

g. **Information Dissemination**: Ensure timely and transparent dissemination of information to community members about development projects, policies, decisions, and outcomes. Use various communication channels, including community radio, posters, pamphlets, mobile phones, and social media, to reach diverse audiences and facilitate two-way communication. Accessible and understandable information fosters informed decision-making and promotes accountability and transparency in development initiatives.

h. **Participatory Monitoring and Evaluation (PM&E):** Implement participatory monitoring and evaluation processes that involve community members in assessing project progress, outcomes, and impacts. PM&E methods, such as participatory impact assessments, community scorecards, and outcome mapping, enable communities to track changes, learn from experiences, and adapt strategies based on feedback from beneficiaries and stakeholders.

By employing these community engagement strategies, stakeholders can foster meaningful participation, build trust, and promote collective action within watershed communities, leading to more inclusive, responsive, and sustainable development outcomes.

Capacity Building: Discussing approaches for enhancing the capacity of community members, including training programs, skill development workshops, and knowledge exchange activities. Discussing approaches for enhancing the capacity of community members is crucial for empowering individuals and fostering sustainable development within watershed areas. Here are key approaches for capacity building:

a. **Training Programs**: Implement training programs tailored to the specific needs and priorities of community members within watershed communities. Training programs may cover a wide range of topics, including sustainable agriculture practices, water management techniques, disaster preparedness, natural resource conservation, entrepreneurship skills, gender mainstreaming, and leadership development. Training sessions should be interactive, participatory, and hands-on, incorporating experiential learning methods and real-life case studies to enhance effectiveness and relevance.

b. **Skill Development Workshops**: Organize skill development workshops that provide practical training and hands-on experience in areas such as organic farming, agroforestry, soil conservation, irrigation techniques, animal husbandry, artisanal crafts, renewable energy technologies, and small-scale enterprise development. Skill development workshops equip community members with the technical skills, tools, and knowledge needed to improve their livelihoods, enhance productivity, and adapt to changing environmental conditions.

c. **Knowledge Exchange Activities**: Facilitate knowledge exchange activities that enable community members to share experiences, lessons learned, and best practices with peers, experts, and practitioners. Knowledge exchange activities may include study tours, field visits, farmer field schools, demonstration plots, community learning exchanges, and peer-to-peer mentoring programs. By promoting dialogue, collaboration, and mutual learning, knowledge exchange activities foster innovation, build social capital, and strengthen collective action within watershed communities.

d. **Participatory Learning Methods**: Adopt participatory learning methods that engage community members as active participants in the learning process. Participatory learning methods, such as role-plays, group discussions, games, storytelling, and participatory videos, empower participants to co-create knowledge, articulate their perspectives, and contribute their insights to the learning process. Participatory learning methods promote inclusivity, ownership, and empowerment, enabling community members to take ownership of their own development.

e. **Tailored Approaches**: Customize capacity building approaches to the socio-economic, cultural, and educational context of the target community. Recognize diversity within communities and ensure that capacity building interventions are inclusive, culturally sensitive, and gender-responsive. Tailored approaches may involve adapting training materials, methodologies, and delivery formats to meet the specific needs and preferences of different groups within the community, including women, youth, elders, and marginalized populations.

f. **Continuous Learning and Feedback**: Foster a culture of continuous learning and feedback by providing ongoing support, mentoring, and follow-up to community members participating in capacity building activities. Encourage participants to apply newly acquired knowledge and skills in their daily lives, monitor progress, and provide opportunities for reflection, self-assessment, and peer-to-peer support. Continuous learning and feedback mechanisms enable capacity building initiatives to be adaptive, responsive, and relevant to evolving community needs and priorities.

By adopting these approaches for enhancing the capacity of community members within watershed areas, stakeholders can empower individuals, strengthen local institutions, and promote sustainable development outcomes that improve the well-being and resilience of communities over the long term.

Empowerment and Ownership: Examining the importance of empowering communities to take ownership of watershed management initiatives, including rights-based approaches, institutional strengthening, and decentralization of authority. Examining the importance of empowering communities to take ownership of watershed management initiatives in India underscores the significance of rights-based approaches, institutional strengthening, and decentralization of authority. Here's how these factors contribute to empowerment and ownership within watershed management:

a. **Rights-Based Approaches**: Rights-based approaches recognize the entitlements of communities to access and manage natural resources, including land, water, forests, and biodiversity, based on principles of equity, justice, and participation. Empowering communities to assert their rights to land tenure, water resources, and customary use rights strengthens their ownership and stewardship of watershed areas. Rights-based approaches promote inclusive decision-making processes, ensure accountability of duty bearers, and address underlying power imbalances that marginalize vulnerable groups within communities.

b. **Institutional Strengthening**: Strengthening local institutions, such as community-based organizations, watershed committees, user groups, and village councils, enhances their capacity to govern natural resources, mobilize collective action, and sustainably manage watershed areas. Empowering local institutions involves building their technical, managerial, and organizational capacities through training, mentoring, and institutional support. Strengthened institutions serve as platforms for community participation, knowledge sharing, and collaborative decision-making, fostering a sense of ownership and responsibility for watershed management.

c. **Decentralization of Authority:** Decentralization of authority involves devolving decision-making powers, resources, and responsibilities from higher levels of government to local communities and institutions. Decentralized governance structures, such as gram panchayats, village councils, and watershed committees, enable communities to participate in planning, implementing, and monitoring watershed management initiatives. Empowering communities with decision-making authority over local resources enhances their ownership, agency, and accountability in managing watersheds according to their needs, priorities, and aspirations.

d. **Community Participation and Engagement**: Actively involving communities in all stages of watershed management, from problem identification to project evaluation, fosters a sense of ownership and commitment to sustainable outcomes. Community participation mechanisms, such as participatory planning processes, community consultations, and stakeholder engagement forums, provide opportunities for communities to voice their concerns, contribute local knowledge, and co-design interventions that are contextually appropriate and socially inclusive.

e. **Capacity Building and Empowerment**: Investing in capacity building programs that enhance the knowledge, skills, and leadership capabilities of community members strengthens their ability to engage effectively in watershed management initiatives. Empowerment through education, training, and skill development enables communities to advocate for their interests, negotiate with external stakeholders, and mobilize resources for collective action. Capacity building fosters a sense of ownership, self-reliance, and resilience within communities, empowering them to lead their own development processes.

f. **Partnership and Collaboration**: Collaborative partnerships between government agencies, civil society organizations, research institutions, private sector actors, and local communities are essential for empowering communities and fostering ownership of watershed management initiatives. Partnerships built on mutual respect, trust, and shared goals facilitate knowledge sharing, resource mobilization, and collective action toward common objectives. Engaging in partnerships empowers communities to leverage external expertise, resources, and networks while retaining control over decision-making processes and project implementation.

Examining the importance of empowering communities to take ownership of watershed management initiatives in India highlights the transformative potential of rights-based approaches, institutional strengthening, and decentralization of authority in fostering sustainable development, social equity, and environmental resilience within watershed areas. Empowered communities are better equipped to address local challenges, harness opportunities, and realize their vision for a more prosperous and sustainable future.

5.2 Multi-Stakeholder Collaboration and Partnership Building

Table: Table illustrates the diverse roles and contributions of various stakeholder groups in watershed management, along with examples of collaboration activities and partnerships that facilitate multi-stakeholder engagement in sustainable watershed management initiatives.

S. No.	Stakeholder Group	Role and Contribution	Collaboration Activities	Examples of Partnerships
1.	Government Agencies	Policy formulation and regulation enforcement	- Developing watershed management plans	- Ministry of Environment, Forests and Climate Change
			- Implementing regulatory measures	- State Departments of Water Resources
			- Providing funding and resources for projects	- National and State Disaster Management Authorities
			- Conducting research and monitoring programs	
2.	Local Communities	On-the-ground knowledge and participation	- Participating in decision-making processes	- Community-based organizations
			Implementing conservation and restoration projects	- Indigenous communities
			- Monitoring environmental changes	- Farmer cooperatives
			- Providing labor and resources for initiatives	
3.	NGOs and Civil Society	Advocacy and community engagement	- Raising awareness about watershed issues	- Environmental NGOs
			- Mobilizing resources and volunteers	- Wildlife conservation organizations
			- Capacity building and training programs	- Women's and youth groups
			- Implementing pilot projects	
4.	Academic Institutions	Research and technical expertise	- Conducting studies on watershed dynamics	- Universities and research institutes

			- Developing innovative solutions	- Research centers focused on water resources management
			- Providing training and education	
			- Collaborating on data collection and analysis	
5.	Private Sector	Investment and technology innovation	- Investing in sustainable development initiatives	- Water utilities and infrastructure companies
			- Developing and deploying technology solutions	- Agribusiness corporations
			- Supporting corporate social responsibility efforts	- Renewable energy companies
			- Partnering on watershed conservation projects	

Stakeholder Identification: Identifying key stakeholders and actors involved in watershed management, including government agencies, NGOs, private sector entities, research institutions, and local communities. Identifying key stakeholders and actors involved in watershed management is essential for fostering collaboration, coordination, and effective decision-making. Here are the main stakeholders and actors typically involved in watershed management:

a. **Government Agencies:** Government agencies at various levels, including national, state, and local authorities, play a significant role in watershed management. These may include ministries of environment, water resources, agriculture, rural development, forestry, and local governance bodies such as gram panchayats, municipal corporations, and district administrations. Government agencies formulate policies, allocate resources, regulate land and water use, and implement watershed management programs and projects.

b. **Non-Governmental Organizations (NGOs)**: NGOs play a vital role in implementing watershed management initiatives, providing technical assistance, capacity building, and community mobilization support. NGOs often work closely with local communities, government agencies, and other stakeholders to address environmental challenges, promote sustainable livelihoods, and build resilience within watershed areas. Examples of NGOs involved in watershed management include environmental conservation organizations, community development NGOs, and advocacy groups.

c. **Private Sector Entities**: Private sector entities, including agricultural corporations, agribusinesses, mining companies, and industries, have a stake in watershed management due to their reliance on natural resources and water for production processes. Private sector engagement in watershed management may involve adopting sustainable practices, investing in watershed restoration projects, and supporting community-based initiatives that promote environmental conservation and responsible resource use.

d. **Research Institutions**: Research institutions, including universities, research centers, and scientific organizations, contribute to watershed management through applied research, data collection, analysis, and knowledge generation. Research institutions conduct studies on hydrology, soil erosion, biodiversity, climate change impacts, and socio-economic dynamics within watershed areas. They provide evidence-based recommendations, technical expertise, and innovative solutions to inform policy and practice in watershed management.

e. **Local Communities:** Local communities living within watershed areas are key stakeholders in watershed management, as they are directly affected by changes in land and water resources. Local communities rely on watershed ecosystems for their livelihoods, food security, water supply, and cultural identity. Community participation and engagement are essential for identifying local priorities, building consensus, and ensuring the sustainability of watershed management interventions.

f. **Academic Institutions:** Academic institutions, including schools, colleges, and training centers, contribute to watershed management through education, awareness raising, and capacity building initiatives. Academic institutions offer academic programs, courses, and workshops on topics related to watershed management, environmental science, and sustainable development. They play a role in educating future professionals, practitioners, and policymakers in the principles and practices of watershed management.

g. **Civil Society Organizations:** Civil society organizations, including community-based organizations, advocacy groups, and social movements, advocate for the rights and interests of marginalized communities and promote participatory approaches to watershed management. Civil society organizations mobilize communities, raise awareness about environmental issues, and hold government and private sector actors accountable for their actions. They play a crucial role in facilitating community empowerment, social mobilization, and grassroots activism within watershed areas.

Identifying key stakeholders and actors involved in watershed management requires a participatory and inclusive approach that recognizes the diverse interests, perspectives, and contributions of different stakeholders. By engaging with stakeholders collaboratively, sharing information transparently, and fostering dialogue and cooperation, stakeholders can work together to address complex challenges and achieve sustainable outcomes within watershed areas.

Collaborative Platforms: Discussing mechanisms for fostering collaboration and partnership among stakeholders, such as watershed committees, multi-stakeholder platforms, and public-private partnerships. Discussing mechanisms for fostering collaboration and partnership among stakeholders in watershed management is crucial for promoting shared responsibility, leveraging resources, and achieving collective impact. Here are key collaborative platforms commonly used to facilitate collaboration among stakeholders:

a. **Watershed Committees:** Watershed committees are multi-stakeholder bodies responsible for coordinating and overseeing watershed management activities at the local level. Watershed committees typically comprise representatives from government agencies, NGOs, community-based organizations, local communities, and other relevant stakeholders. These committees serve as forums for collaborative decision-making, planning, and implementation of watershed management initiatives. Watershed committees promote inclusivity, transparency, and accountability in governance processes and help build consensus among diverse stakeholders.

b. **Multi-Stakeholder Platforms (MSPs):** Multi-stakeholder platforms bring together a wide range of actors, including government agencies, civil society organizations, private sector entities, research institutions, and local communities, to address complex issues related to watershed management. MSPs provide neutral spaces for dialogue, knowledge sharing, and joint problem-solving, fostering mutual understanding, trust, and cooperation among stakeholders. MSPs facilitate the exchange of expertise, resources, and experiences, leading to more integrated and sustainable approaches to watershed management.

c. **Public-Private Partnerships (PPPs):** Public-private partnerships involve collaboration between government agencies and private sector entities to jointly plan, finance, and implement watershed management initiatives. PPPs leverage the strengths and resources of both sectors to address common challenges and achieve shared goals. Private sector involvement in watershed management may include investments in sustainable agriculture, water infrastructure, eco-tourism, and corporate social responsibility initiatives that contribute to environmental conservation and community development. PPPs promote innovation, efficiency, and accountability in delivering watershed management services and outcomes.

d. **Community-Based Organizations (CBOs):** Community-based organizations, such as self-help groups, cooperatives, and user associations, play a crucial role in facilitating collaboration and partnership among local communities and external stakeholders. CBOs serve as intermediaries between communities and government agencies, NGOs, and other actors, advocating for community interests, mobilizing resources, and implementing grassroots projects. By strengthening the capacity of CBOs, stakeholders can empower communities to participate more effectively in collaborative initiatives and take ownership of watershed management processes.

e. **Learning Networks and Knowledge Platforms**: Learning networks and knowledge platforms provide opportunities for stakeholders to share lessons learned, best practices, and innovative solutions in watershed management. These platforms may include online forums, workshops, conferences, and knowledge exchange events that promote cross-sectoral collaboration, peer learning, and continuous improvement. Learning networks facilitate the dissemination of evidence-based information, promote dialogue among diverse stakeholders, and catalyze collective action toward common objectives in watershed management.

f. **Advisory Groups and Technical Committees**: Advisory groups and technical committees composed of subject matter experts, scientists, practitioners, and community representatives provide technical guidance, strategic advice, and oversight for watershed management initiatives. These groups may advise on policy formulation, project design, research priorities, and monitoring and evaluation frameworks. Advisory groups and technical committees ensure that watershed management interventions are informed by scientific evidence, technical expertise, and local knowledge, enhancing their relevance, effectiveness, and sustainability.

By leveraging collaborative platforms such as watershed committees, multi-stakeholder platforms, public-private partnerships, and community-based organizations, stakeholders can harness the collective expertise, resources, and commitment of diverse actors to address complex challenges and advance sustainable watershed management initiatives. Collaborative approaches foster inclusive decision-making, build social capital, and promote synergy among stakeholders, leading to more resilient, equitable, and environmentally sustainable outcomes within watershed areas.

Conflict Resolution: Addressing challenges related to conflicting interests, power dynamics, and resource competition among stakeholders through dialogue, negotiation, and conflict resolution mechanisms. Addressing challenges related to conflicting interests, power dynamics, and resource competition among stakeholders in watershed management in India requires effective conflict resolution mechanisms. Here are key strategies for managing conflicts:

a. **Dialogue and Negotiation**: Encourage stakeholders to engage in open and constructive dialogue to identify underlying concerns, interests, and priorities. Facilitate negotiations among conflicting parties to explore mutually acceptable solutions and compromises. Dialogue and negotiation processes should be inclusive, transparent, and facilitated by neutral mediators to build trust and foster cooperation among stakeholders.

b. **Mediation and Facilitation**: Utilize mediation and facilitation techniques to help parties in conflict find common ground and resolve differences amicably. Trained mediators or facilitators can assist stakeholders in exploring options, generating creative solutions, and reaching agreements that address the interests of all parties. Mediation and facilitation processes promote active listening, empathy, and understanding, leading to sustainable resolutions of conflicts.

c. **Conflict Mapping and Analysis**: Conduct conflict mapping and analysis to identify the root causes, drivers, and dynamics of conflicts within watershed areas. Analyze socio-economic, political, and environmental factors that contribute to conflicts over land, water, and natural resources. Understanding the complexity of conflicts enables stakeholders to develop targeted interventions and preventive measures to address underlying grievances and prevent escalation.

d. **Participatory Decision-Making**: Promote participatory decision-making processes that involve all stakeholders in identifying issues, exploring options, and making informed decisions collectively. Ensure that marginalized groups, such as women, indigenous communities, and landless farmers, have a voice in decision-making processes to ensure their interests are represented and considered. Participatory decision-making builds ownership, legitimacy, and accountability in watershed management initiatives.

e. **Legal and Policy Frameworks**: Establish clear legal and policy frameworks governing rights, responsibilities, and resource allocation within watershed areas. Ensure that laws and regulations related to land tenure, water rights, environmental protection, and community participation are enforced and respected. Legal frameworks provide a basis for resolving disputes, protecting vulnerable groups, and promoting equity and justice in watershed management.

f. **Community-Based Conflict Resolution Mechanisms**: Strengthen community-based conflict resolution mechanisms, such as traditional dispute resolution mechanisms, village councils, and community mediation committees. Empower local communities to resolve conflicts at the grassroots level using culturally appropriate and locally accepted methods. Community-based conflict resolution mechanisms build social cohesion, promote reconciliation, and reduce the need for external intervention.

g. **Capacity Building for Conflict Management**: Provide training and capacity building programs to stakeholders, including government officials, community leaders, civil society representatives, and private sector actors, in conflict management skills and techniques. Enhance stakeholders' abilities to identify, prevent, and resolve conflicts through communication, negotiation, mediation, and consensus-building strategies. Capacity building fosters a culture of conflict resolution and collaboration, strengthening the resilience of watershed communities to manage conflicts effectively.

h. **Monitoring and Evaluation:** Establish mechanisms for monitoring and evaluating conflict resolution processes and outcomes to assess their effectiveness and identify lessons learned. Monitor changes in stakeholder relationships, levels of cooperation, and the resolution of conflicts over time. Evaluate the impacts of conflict resolution interventions on social cohesion, resource management, and sustainable development within watershed areas. Monitoring and evaluation provide valuable feedback for refining conflict resolution strategies and improving future interventions.

By addressing challenges related to conflicting interests, power dynamics, and resource competition through dialogue, negotiation, and conflict resolution mechanisms, stakeholders can promote harmony, cooperation, and equitable resource allocation within watershed areas in India. Effective conflict resolution contributes to social stability, environmental sustainability, and inclusive development, enabling stakeholders to work together toward shared goals and aspirations.

5.3 Planning and Implementation of Watershed Development Projects

Integrated Planning Processes: Describing the steps involved in watershed planning, including situation analysis, goal setting, strategy formulation, action planning, and resource allocation. Describing the steps involved in integrated watershed planning in India involves understanding the complexities of the process and the unique context of the country's diverse landscapes and socio-economic conditions. Here are the key steps typically involved in watershed planning:

a. **Situation Analysis**: Conduct a comprehensive assessment of the biophysical, socio-economic, and institutional factors shaping watershed dynamics. This involves collecting and analyzing data on land use, land cover, soil types, topography, hydrology, climate patterns, demographic characteristics, livelihood strategies, resource use patterns, infrastructure, governance structures, and existing policies and programs related to watershed management. Situation analysis provides a baseline understanding of the current status of the watershed and identifies key challenges, opportunities, and drivers of change.

b. **Stakeholder Engagement**: Engage stakeholders representing diverse interests, including government agencies, local communities, NGOs, private sector entities, research institutions, and civil society organizations, in the planning process. Conduct stakeholder consultations, focus group discussions, workshops, and participatory mapping exercises to gather input, feedback, and local knowledge. Stakeholder engagement ensures that planning processes are inclusive, transparent, and reflective of the needs and priorities of all relevant actors within the watershed.

c. **Goal Setting**: Define clear and achievable goals and objectives for watershed management based on the findings of the situation analysis and stakeholder consultations. Goals may include enhancing water security, improving soil health, increasing agricultural productivity, reducing erosion and sedimentation, conserving biodiversity, mitigating climate change impacts, and enhancing livelihood opportunities for local communities. Goal setting provides a framework for guiding planning efforts and measuring progress toward desired outcomes.

d. **Strategy Formulation**: Develop strategies and interventions to achieve the identified goals and objectives of watershed management. Strategies may encompass a mix of technical, institutional, and policy measures tailored to the specific needs and characteristics of the watershed. Examples of strategies include promoting sustainable land use practices, implementing soil and water conservation measures, rehabilitating degraded ecosystems, enhancing water harvesting and storage capacities, promoting alternative livelihood options, strengthening institutional capacities, and fostering community participation in decision-making processes. Strategy formulation involves prioritizing interventions based on their feasibility, effectiveness, and potential impact on watershed sustainability.

e. **Action Planning:** Develop detailed action plans outlining specific activities, timelines, responsibilities, and resource requirements for implementing the selected strategies. Action plans specify the sequence of activities, milestones, and indicators for monitoring progress and measuring success. Action planning involves breaking down broad strategies into manageable tasks and identifying the necessary inputs, technical assistance, funding sources, and partnerships required to implement them effectively. Action plans serve as roadmaps for guiding implementation efforts and ensuring accountability among stakeholders.

f. **Resource Allocation**: Allocate financial, human, and technical resources to support the implementation of watershed management activities outlined in the action plans. Identify funding sources, such as government budgets, grants, loans, user fees, and private sector investments, to finance priority interventions. Mobilize technical expertise, equipment, and materials needed to execute planned activities. Resource allocation involves coordinating efforts among different stakeholders, leveraging existing resources, and exploring innovative financing mechanisms to maximize the impact of watershed management investments.

By following these steps in integrated watershed planning, stakeholders can develop comprehensive and contextually appropriate strategies for sustainable watershed management in India. Integrated planning processes promote collaboration, coordination, and synergy among stakeholders, leading to more effective use of resources, better management of natural resources, and improved livelihoods for communities dependent on watershed ecosystems.

Project Design and Management: Discussing principles of project design, including setting SMART (Specific, Measurable, Achievable, Relevant, Time-bound) objectives, developing work plans, and establishing monitoring and evaluation frameworks.

Discussing the principles of project design, including setting SMART objectives, developing work plans, and establishing monitoring and evaluation frameworks, is crucial for ensuring the effectiveness and success of watershed management projects in India. Here's how these principles can be applied, along with an Indian example:

a. **Setting SMART Objectives**

- **Specific:** Objectives should be clearly defined and focused on addressing specific challenges or achieving desired outcomes within the watershed context. For example, an objective could be to reduce soil erosion by implementing contour bunding and agroforestry practices in a degraded watershed area.
- **Measurable**: Objectives should be quantifiable, allowing progress to be tracked and measured over time. For instance, the objective to increase groundwater recharge could specify a measurable target, such as raising the water table by a certain number of meters within a specified timeframe.
- **Achievable:** Objectives should be realistic and feasible given the available resources, capacities, and constraints. It's essential to

consider socio-economic, environmental, and institutional factors that may influence the achievability of objectives.

- **Relevant:** Objectives should be relevant to the overall goals of watershed management, addressing priority issues and contributing to sustainable development outcomes. For example, an objective related to improving agricultural productivity aligns with the broader goal of enhancing food security and livelihoods in the watershed.
- **Time-bound:** Objectives should have clear timelines and deadlines for achievement, helping to create a sense of urgency and accountability. Establishing time-bound targets ensures that progress is monitored and adjustments can be made as needed. For instance, a project objective could be to rehabilitate a degraded watershed area within three years.

b. Developing Work Plans

- Once SMART objectives are defined, develop detailed work plans outlining the activities, tasks, responsibilities, timelines, and resources required to achieve each objective. Work plans provide a roadmap for project implementation, guiding the sequence of activities and ensuring efficient use of resources.
- Work plans should be developed collaboratively with stakeholders, drawing on their expertise, local knowledge, and perspectives. Stakeholder involvement in work plan development fosters ownership, buy-in, and commitment to project goals and activities.
- Work plans should be flexible and adaptable to changing circumstances, allowing for adjustments based on evolving needs, priorities, and feedback from stakeholders. Regular reviews and updates to work plans help to ensure that project activities remain relevant and responsive to the dynamic context of watershed management.

c. Establishing Monitoring and Evaluation Frameworks

- Design monitoring and evaluation (M&E) frameworks to assess the progress, effectiveness, and impact of watershed management interventions against the defined objectives and targets.
- Develop indicators and data collection methods to measure changes in key parameters related to watershed health, environmental sustainability, livelihoods, and socio-economic well-being. Indicators should be aligned with SMART objectives and be feasible to measure with available resources.

- Establish baseline data to provide a reference point for measuring change and progress over time. Baseline data collection should occur before project implementation begins and include relevant socio-economic, biophysical, and institutional indicators.
- Implement regular monitoring activities to track progress, identify challenges, and capture lessons learned throughout the project lifecycle. Monitoring activities may include field surveys, participatory assessments, remote sensing, and data analysis.
- Conduct periodic evaluations to assess the overall effectiveness, efficiency, and impact of watershed management interventions. Evaluations should be participatory, transparent, and rigorous, involving stakeholders in data collection, analysis, and interpretation.
- Use M&E findings to inform decision-making, adaptive management, and project refinements. Share evaluation results with stakeholders to promote learning, accountability, and continuous improvement in watershed management practices.

Example

In the Indian context, the Integrated Watershed Management Programme (IWMP) implemented by the Ministry of Rural Development provides a relevant example of project design and management principles in watershed development. The IWMP aims to restore ecological balance by harnessing natural resources and improving livelihoods in rainfed areas through community-based watershed management approaches.

Setting SMART Objectives: In IWMP projects, SMART objectives are defined to address specific challenges such as soil erosion, water scarcity, and declining agricultural productivity in targeted watershed areas. For example, an objective could be to increase groundwater recharge by implementing rainwater harvesting structures and reforestation measures within a specified timeframe.

Developing Work Plans: IWMP projects develop detailed work plans outlining activities such as contour bunding, check dams construction, afforestation, soil conservation, and livelihood diversification initiatives. Work plans specify the roles and responsibilities of different stakeholders, allocate resources, and establish timelines for project implementation.

Establishing Monitoring and Evaluation Frameworks: IWMP projects establish monitoring and evaluation frameworks to track progress, measure outcomes, and assess the impact of watershed interventions. Indicators related to soil moisture levels, vegetation cover, crop yields, water availability, and income generation are monitored regularly to evaluate the effectiveness of project interventions. Evaluations are conducted periodically to review project performance, identify lessons learned, and inform adaptive management decisions.

By applying these project design and management principles, watershed management projects in India can achieve their objectives effectively, maximize impact, and contribute to sustainable development outcomes in rural areas.

Technical Interventions: Exploring a range of technical interventions and measures used in watershed development projects, including soil conservation, water harvesting, afforestation, agroforestry, and sustainable agriculture practices. Exploring a range of technical interventions and measures used in watershed development projects reveals the diversity of approaches aimed at improving watershed health, enhancing ecosystem services, and supporting sustainable livelihoods. Here are some key technical interventions commonly employed in watershed development projects:

a. **Soil Conservation**
    - **Contour Bunding:** Constructing bunds along the contour lines of sloping lands to reduce soil erosion and retain water.
    - **Terracing:** Building terraces or steps on hillsides to reduce soil erosion, increase water infiltration, and improve land stability.
    - **Vegetative Measures:** Establishing vegetative barriers such as grass strips, cover crops, and agroforestry systems to protect soil from erosion by wind and water.
    - **Conservation Tillage:** Adopting minimum tillage or zero tillage practices to reduce soil disturbance, improve soil structure, and minimize erosion.

b. **Water Harvesting**
    - **Check Dams:** Constructing small earthen or stone dams across ephemeral streams to impound water, recharge groundwater, and prevent soil erosion.
    - **Contour Trenches**: Digging trenches along contour lines to capture and slow down surface runoff, allowing water to infiltrate into the soil and recharge aquifers.
    - **Rooftop Rainwater Harvesting**: Installing rainwater harvesting structures on rooftops to capture rainwater for domestic use, livestock watering, and groundwater recharge.

- Percolation Ponds: Excavating shallow depressions in the landscape to collect and store runoff water, allowing it to percolate into the ground and recharge aquifers.

c. **Afforestation and Reforestation**

- **Planting Trees and Shrubs**: Establishing tree and shrub plantations on degraded lands to stabilize soil, enhance biodiversity, and sequester carbon.
- **Agroforestry Systems:** Integrating trees with agricultural crops and livestock to improve soil fertility, provide shade, regulate microclimates, and diversify livelihood options.
- **Social Forestry Initiatives**: Involving local communities in tree planting activities on common lands, road verges, and degraded areas to promote community stewardship of natural resources.

d. **Sustainable Agriculture Practices**

- **Crop Diversification**: Introducing diverse cropping systems that incorporate a variety of crops, including legumes, grains, vegetables, and fruits, to enhance soil fertility, pest management, and resilience to climate variability.
- **Organic Farming**: Adopting organic farming practices that avoid synthetic inputs and prioritize soil health, biodiversity conservation, and ecosystem resilience.
- **Integrated Pest Management (IPM):** Implementing IPM strategies that combine cultural, biological, and ecological methods to manage pests and diseases while minimizing chemical pesticide use.
- **Conservation Agriculture**: Emphasizing minimum soil disturbance, permanent soil cover, and crop rotation to improve soil structure, water retention, and crop productivity.

e. **Livestock Management**

- **Grazing Management**: Implementing rotational grazing systems, fodder banks, and stall-feeding practices to prevent overgrazing, restore vegetation cover, and improve rangeland productivity.
- **Stall-fed Livestock**: Encouraging stall-feeding of livestock with crop residues, agroforestry products, and fodder crops to reduce pressure on natural pastures and minimize soil erosion.
- **Livestock Integration**: Integrating livestock with cropping systems through manure application, bio-gas production, and nutrient cycling to improve soil fertility and enhance farm productivity.

f. **Water Management**

- **Micro-Irrigation Systems**: Installing drip irrigation, sprinkler irrigation, and other micro-irrigation systems to optimize water use efficiency, reduce water wastage, and enhance crop yields.
- **Water Budgeting**: Implementing water budgeting techniques to assess water availability, demand, and use patterns, enabling farmers to make informed decisions on water allocation and management.
- **Water Quality Management**: Implementing measures to reduce water pollution from agricultural runoff, livestock waste, and other sources, including vegetative buffer strips, riparian vegetation, and constructed wetlands.

These technical interventions are often integrated into holistic watershed management approaches that consider the interconnections between soil, water, vegetation, and socio-economic systems. By implementing a combination of these interventions tailored to the specific conditions and needs of each watershed, stakeholders can improve watershed health, enhance ecosystem resilience, and promote sustainable livelihoods for local communities.

5.4 Monitoring, Evaluation and Adaptive Management Practices

Monitoring Indicators: Identifying key performance indicators (KPIs) and metrics for measuring the effectiveness of watershed management interventions, including ecological, social, and economic indicators. Identifying key performance indicators (KPIs) and metrics for measuring the effectiveness of watershed management interventions is crucial for assessing progress, guiding decision-making, and ensuring accountability. Here are some examples of KPIs and metrics categorized into ecological, social, and economic indicators:

a. **Ecological Indicators**

- **Soil Erosion Rate**: Measure changes in soil erosion rates using techniques such as erosion pins, sediment traps, or remote sensing imagery to assess the effectiveness of soil conservation measures.
- **Vegetation Cover:** Monitor changes in vegetation cover using satellite imagery, aerial photography, or ground-based surveys to evaluate the success of afforestation and reforestation efforts.
- **Water Quality**: Assess water quality parameters such as turbidity, pH, dissolved oxygen, and nutrient concentrations to evaluate the impact of watershed management interventions on water quality.
- **Biodiversity Index**: Calculate biodiversity indices based on species

richness, abundance, and diversity metrics to monitor changes in ecosystem health and habitat quality.

b. **Social Indicators**

- **Community Participation**: Measure levels of community participation and engagement in watershed management activities through surveys, focus group discussions, or participatory mapping exercises to gauge local ownership and empowerment.
- **Livelihood Diversification**: Assess changes in livelihood strategies, income sources, and employment opportunities among watershed communities to determine the socio-economic impacts of watershed interventions.
- **Social Equity:** Evaluate the distribution of benefits and costs of watershed management interventions among different social groups, including marginalized communities, women, and landless farmers.
- **Capacity Building:** Track indicators related to the uptake of training programs, skill development workshops, and knowledge exchange activities to assess the capacity building efforts and empowerment of stakeholders.

c. **Economic Indicators**

- **Cost-effectiveness**: Calculate the cost per unit of outcome achieved, such as cost per hectare of land restored or cost per cubic meter of water harvested, to assess the cost-effectiveness of watershed management interventions.
- **Income Generation**: Measure changes in household incomes, agricultural yields, and asset ownership among watershed communities to evaluate the economic benefits of improved land and water management practices.
- **Return on Investment (ROI):** Estimate the return on investment from watershed management interventions by comparing the economic benefits generated, such as increased agricultural productivity or reduced flood damage, to the costs incurred.
- **Market Access**: Assess changes in market access, value chains, and market linkages for agricultural products and non-timber forest products (NTFPs) to evaluate the impact of watershed interventions on local economies.

These indicators can be monitored and tracked over time using quantitative and qualitative data collection methods, including field surveys, interviews, focus group discussions, participatory assessments, and secondary data analysis. By monitoring a combination of ecological, social, and economic indicators, stakeholders can gain a comprehensive understanding of the outcomes and impacts of watershed management interventions, informing adaptive management decisions and guiding future investments in sustainable watershed development.

Evaluation Methods: Discussing methods and tools for conducting baseline surveys, impact assessments, and midterm and final evaluations of watershed projects, including participatory evaluation approaches. Discussing methods and tools for conducting baseline surveys, impact assessments, and midterm and final evaluations of watershed projects, including participatory evaluation approaches, provides insights into comprehensive evaluation processes that capture the complexities of watershed management interventions. Here are some methods and tools commonly used in each phase of project evaluation:

a. **Baseline Surveys**

- **Quantitative Surveys:** Administer structured questionnaires or surveys to collect baseline data on socio-economic indicators, land use practices, natural resource conditions, and community perceptions and priorities.
- **Key Informant Interviews:** Conduct interviews with key informants, such as community leaders, government officials, and local experts, to gather qualitative insights and contextual information relevant to the project area.
- **Focus Group Discussions (FGDs):** Organize FGDs with diverse stakeholder groups to explore perceptions, knowledge, attitudes, and aspirations related to watershed management, facilitating participatory data collection and community engagement.
- **Participatory Mapping:** Engage communities in participatory mapping exercises to spatially visualize resource use patterns, land tenure arrangements, water sources, and ecosystem services, providing valuable insights into local dynamics and vulnerabilities.

b. **Impact Assessments**

- **Quantitative Surveys**: Administer follow-up surveys or questionnaires to assess changes in key indicators and outcomes over time, comparing post-intervention data with baseline data to measure project impacts.
- **Remote Sensing and GIS Analysis**: Use remote sensing imagery and geographic information systems (GIS) to analyze changes in land cover, vegetation density, soil erosion, and water availability at different spatial and temporal scales.

- **Economic Analysis:** Conduct cost-benefit analysis, cost-effectiveness analysis, or return on investment (ROI) calculations to evaluate the economic viability and efficiency of watershed management interventions.
- **Ecological Monitoring**: Implement field-based ecological monitoring programs to track changes in ecological indicators such as biodiversity, soil health, water quality, and hydrological processes, providing insights into ecosystem responses to project interventions.

c. Midterm and Final Evaluations

- **Mixed-Methods Approaches:** Combine quantitative and qualitative methods to conduct comprehensive midterm and final evaluations, triangulating data from surveys, interviews, focus groups, field observations, and secondary sources.
- **Outcome Mapping:** Use outcome mapping techniques to identify and visualize changes in behaviors, relationships, and capacities among stakeholders, mapping pathways of change and contribution analysis.
- **Participatory Action Research (PAR):** Collaborate with stakeholders in PAR processes that engage communities in co-designing, implementing, and evaluating project interventions, promoting local ownership, learning, and adaptive management.
- **Stakeholder Workshops:** Organize stakeholder workshops or reflection sessions to review evaluation findings, validate results, and co-generate recommendations for project adjustments and future planning.

d. Participatory Evaluation Approaches

- **Most Significant Change (MSC):** Facilitate MSC processes where stakeholders identify and discuss significant changes observed as a result of the project, promoting dialogue, reflection, and collective sense-making.
- **Community Scorecards**: Develop community scorecards or performance dashboards that allow stakeholders to assess project progress, outcomes, and impacts against predefined criteria and indicators, fostering transparency and accountability.
- **Participatory Video:** Empower communities to produce participatory videos documenting their experiences, challenges, and successes with watershed management interventions, facilitating storytelling, advocacy, and learning.

By employing a combination of these methods and tools, watershed projects can generate robust evidence, capture diverse perspectives, and promote stakeholder engagement throughout the evaluation process. Participatory evaluation approaches enhance the relevance, credibility, and utilization of evaluation findings, empowering communities to drive positive change and sustainably manage their watershed resources.

Adaptive Management Strategies: Exploring strategies for adaptive management, including learning-by-doing, feedback loops, adaptive experimentation, and continuous improvement processes. Exploring strategies for adaptive management in watershed projects involves embracing flexibility, learning, and responsiveness to evolving conditions and feedback. Here are some key strategies for adaptive management:

a. **Learning-by-Doing**
   - Encourage project teams and stakeholders to learn from practical experiences, experimentation, and iterative implementation.
   - Promote a culture of reflection, inquiry, and openness to new ideas and insights gained through hands-on engagement in watershed management activities.
   - Emphasize the value of failure as an opportunity for learning and improvement, encouraging experimentation and innovation in finding solutions to complex challenges.

b. **Feedback Loops**
   - Establish mechanisms for capturing, synthesizing, and integrating feedback from diverse stakeholders, including communities, technical experts, government agencies, and civil society organizations.
   - Use participatory monitoring and evaluation approaches to facilitate ongoing dialogue, reflection, and joint sense-making among stakeholders.
   - Regularly review monitoring data, anecdotal evidence, and stakeholder perceptions to identify emerging issues, trends, and unintended consequences of project interventions.

c. **Adaptive Experimentation**
   - Design and implement adaptive experiments to test hypotheses, explore alternative strategies, and assess the effectiveness of innovative approaches in watershed management.
   - Apply a structured approach to experimentation, incorporating control groups, replication, and randomization where feasible to improve the rigor and validity of findings.

- Document and share lessons learned from adaptive experiments, promoting cross-learning and knowledge exchange among project teams and external partners.

d. Continuous Improvement Processes

- Institutionalize mechanisms for continuous reflection, review, and adjustment of project strategies, plans, and actions based on new information, changing circumstances, and stakeholder feedback.
- Conduct regular debriefing sessions, after-action reviews, or project retrospectives to identify successes, challenges, and areas for improvement in project implementation.
- Facilitate collaborative problem-solving and decision-making processes that involve stakeholders in co-designing adaptive responses and refining project approaches.

e. Scenario Planning

- Anticipate and prepare for uncertain future conditions and potential shocks by developing and testing alternative scenarios and response strategies.
- Engage stakeholders in scenario planning exercises to explore plausible future scenarios, assess their potential impacts on watershed systems and communities, and identify robust strategies for building resilience and adaptive capacity.

f. Resilience Assessments

- Conduct resilience assessments to evaluate the adaptive capacity of watershed systems and communities to withstand and recover from disturbances, shocks, and stresses.
- Identify and prioritize interventions that enhance the resilience of social-ecological systems, such as diversifying livelihoods, strengthening social networks, and preserving ecosystem services.

g. Participatory Decision-Making

- Foster participatory decision-making processes that engage stakeholders in co-creating visions, setting priorities, and co-designing adaptive management strategies.
- Empower local communities to take ownership of adaptive management processes, building their capacity to monitor, analyze, and respond to changes in watershed conditions.

By embracing adaptive management strategies, watershed projects can become more responsive, resilient, and effective in achieving their goals of sustainable natural resource management and community development. These strategies enable project teams and stakeholders to navigate uncertainty, complexity, and dynamic change in watershed systems, fostering learning, innovation, and continuous improvement over time.

5.5 Case Studies and Success Stories in Sustainable Watershed Management

Showcasing Best Practices: Highlighting case studies and success stories of integrated watershed management initiatives from different regions, including their objectives, approaches, outcomes, and lessons learned. Highlighting case studies and success stories of integrated watershed management initiatives from different regions provides valuable insights into effective approaches, outcomes, and lessons learned in watershed management. Here are some key elements to include when showcasing best practices:

a. **Objectives**
   - Clearly state the objectives of the watershed management initiative, such as improving water quality, enhancing soil conservation, increasing agricultural productivity, or reducing vulnerability to natural disasters.

b. **Approaches**
   - Describe the approach or strategy employed in the watershed management initiative, including the selection of interventions, stakeholder engagement methods, institutional arrangements, and implementation mechanisms.
   - Highlight any innovative or participatory elements of the approach, such as community-driven decision-making processes, ecosystem-based approaches, or multi-sectoral collaboration.

c. **Outcomes**
   - Summarize the key outcomes and achievements of the watershed management initiative, both in terms of environmental sustainability and socio-economic development.
   - Provide quantitative and qualitative data on changes in ecological indicators, livelihoods, income levels, water availability, and other relevant metrics.
   - Highlight any tangible benefits or improvements observed in the target communities or ecosystems as a result of the initiative.

d. **Lessons Learned**
   - Reflect on the lessons learned from the implementation of the watershed management initiative, including successes, challenges, and opportunities for improvement.

- Identify factors contributing to the success of the initiative, such as strong community engagement, effective governance structures, adaptive management practices, or technical innovation.
- Discuss any barriers or constraints encountered during implementation and strategies employed to overcome them.
- Offer recommendations and insights for scaling up or replicating successful practices in other contexts or regions.

e. Case Studies

- Provide detailed case studies or examples of integrated watershed management initiatives that demonstrate best practices and lessons learned.
- Showcase a diversity of case studies from different geographic regions, ecosystems, and socio-economic contexts to illustrate the versatility and applicability of watershed management approaches.
- Include narratives, testimonials, and visual materials (such as maps, photos, and videos) to bring the case studies to life and engage readers.

By showcasing best practices through case studies and success stories, stakeholders can learn from each other's experiences, replicate successful approaches, and adapt lessons learned to their own watershed management initiatives. These stories can inspire action, build capacity, and catalyze positive change in sustainable natural resource management at local, regional, and global scales.

Scaling Up and Replication: Discussing strategies for scaling up successful watershed management models and replicating best practices in other contexts, including policy support, capacity building, and knowledge sharing mechanisms. Discussing strategies for scaling up successful watershed management models and replicating best practices in other contexts involves addressing various challenges while leveraging opportunities for wider impact. Here are some key strategies to consider:

a. Policy Support

- Advocate for supportive policy frameworks at the national, regional, and local levels that recognize the importance of integrated watershed management and provide incentives for scaling up successful models.
- Collaborate with government agencies, policymakers, and legislators to integrate watershed management principles into relevant policies, plans, and programs related to water resources, land use planning, agriculture, forestry, and rural development.

- Promote policy coherence and alignment across sectors to ensure that watershed management objectives are mainstreamed into broader sustainable development agendas.

b. **Capacity Building**

- Invest in capacity building initiatives to strengthen the technical, institutional, and human capacities of stakeholders involved in watershed management, including government officials, community leaders, civil society organizations, and technical experts.
- Develop training programs, workshops, and knowledge exchange platforms to share lessons learned, best practices, and technical know-how related to integrated watershed management.
- Foster partnerships with academic institutions, research organizations, and training centers to develop curricula, manuals, and educational materials on watershed management for different target audiences.

c. **Knowledge Sharing Mechanisms**

- Establish knowledge sharing networks, communities of practice, and online platforms to facilitate peer-to-peer learning, collaboration, and information exchange among practitioners, policymakers, researchers, and community members.
- Document and disseminate case studies, success stories, and best practices in watershed management through publications, workshops, conferences, webinars, and multimedia channels.
- Promote south-south cooperation and cross-regional learning by facilitating exchange visits, study tours, and twinning arrangements between watershed management initiatives in different regions or countries.

d. **Innovative Financing**

- Explore innovative financing mechanisms, such as payment for ecosystem services (PES), carbon credits, eco-labeling schemes, and public-private partnerships, to mobilize financial resources for scaling up successful watershed management models.
- Engage with philanthropic organizations, impact investors, and corporate partners to leverage private sector investments and corporate social responsibility (CSR) initiatives in watershed conservation and restoration efforts.

- Advocate for increased public investments in watershed management through budget allocations, grants, loans, and subsidies targeted at priority areas and vulnerable communities.

e. Community Empowerment

- Empower local communities to take ownership of scaling up efforts and replicating best practices by building their capacity to plan, implement, and monitor watershed management activities.
- Foster inclusive and participatory decision-making processes that involve communities in identifying priorities, designing interventions, and evaluating outcomes.
- Support the establishment of community-led institutions, such as watershed committees, user groups, and cooperatives, to sustainably manage natural resources and promote collective action for watershed conservation.

By adopting these strategies and approaches, stakeholders can overcome barriers to scaling up successful watershed management models and create enabling conditions for replicating best practices in diverse contexts. Collaboration, innovation, and partnership are essential for achieving widespread adoption and lasting impact in sustainable watershed management at local, national, and global scales.

Innovations and Lessons Learned: Examining innovative approaches, tools, and techniques used in watershed management, as well as challenges faced and lessons learned in implementing sustainable watershed projects. Examining innovative approaches, tools, and techniques used in watershed management provides valuable insights into cutting-edge practices, while also highlighting challenges faced and lessons learned in implementing sustainable watershed projects. Here's a deeper look into these aspects:

a. Innovative Approaches

- **Ecosystem-Based Management**: Emphasizing the protection and restoration of ecosystem services, such as water regulation, soil fertility, and biodiversity conservation, through holistic and integrated approaches.
- **Nature-Based Solutions**: Harnessing the natural functions and processes of ecosystems, such as wetlands, forests, and rivers, to address water-related challenges and enhance resilience to climate change.
- **Green Infrastructure**: Investing in natural and nature-based infrastructure, such as riparian buffers, green roofs, and permeable

pavements, to manage stormwater, reduce flood risk, and improve water quality.

- **Participatory Governance:** Engaging stakeholders in collaborative decision-making processes that empower local communities, enhance social equity, and promote transparent and accountable governance.
- **Technological Innovations**: Leveraging advances in remote sensing, geographic information systems (GIS), hydrological modeling, and sensor technologies to monitor watershed dynamics, predict future scenarios, and inform decision-making.

b. Tools and Techniques

- **Watershed Modeling**: Using hydrological models, such as SWAT (Soil and Water Assessment Tool) or MIKE BASIN, to simulate water flow, sediment transport, and pollutant loading in watershed systems, aiding in watershed planning and management.
- **Decision Support Systems**: Developing decision support tools and software applications that integrate spatial data, ecological models, and socio-economic indicators to assist stakeholders in prioritizing interventions and evaluating trade-offs.
- **Participatory Mapping**: Engaging local communities in participatory mapping exercises, such as participatory GIS (PGIS) or community mapping, to spatially visualize local knowledge, resource use patterns, and priority areas for intervention.
- **Social Network Analysis**: Analyzing social networks and stakeholder interactions using network analysis techniques to understand power dynamics, information flows, and collaboration patterns within watershed governance systems.
- **Payment for Ecosystem Services (PES):** Implementing PES schemes that provide financial incentives to landowners and communities for implementing land management practices that enhance watershed services, such as reforestation, soil conservation, and wetland restoration.

c. Challenges Faced

- **Complexity and Interconnectedness**: Managing watershed systems involves navigating complex interactions between ecological, social, economic, and institutional factors, requiring interdisciplinary approaches and adaptive management strategies.
- **Data Limitations:** Accessing reliable data on watershed characteristics, land use dynamics, hydrological processes, and socio-

economic indicators can be challenging, particularly in data-scarce regions or underfunded contexts.

- **Resource Constraints**: Securing adequate funding, technical expertise, and institutional support for watershed projects may be difficult, especially for grassroots initiatives and community-led efforts.
- **Conflicting Priorities:** Balancing competing interests and conflicting objectives among stakeholders, such as agriculture, industry, conservation, and urban development, requires effective negotiation and conflict resolution mechanisms.

d. Lessons Learned

- **Integrated Approach**: Adopting a holistic and integrated approach to watershed management that considers the interconnections between ecological, social, and economic dimensions is essential for achieving sustainable outcomes.
- **Community Engagement**: Meaningful participation and engagement of local communities in decision-making processes, project design, and implementation are critical for fostering ownership, building social capital, and ensuring project relevance and sustainability.
- **Adaptive Management**: Embracing adaptive management principles, such as learning-by-doing, feedback loops, and continuous improvement processes, enables stakeholders to respond effectively to uncertainty, change, and unforeseen challenges.
- **Knowledge Sharing:** Facilitating knowledge sharing, capacity building, and peer-to-peer learning among stakeholders helps to disseminate best practices, avoid duplication of efforts, and catalyze innovation in watershed management.

By examining innovations and lessons learned in watershed management, stakeholders can identify opportunities for improvement, innovation, and collaboration, ultimately contributing to more effective and sustainable approaches to watershed conservation and management.

Important Questions

1. How do dialogue and negotiation contribute to resolving conflicts among stakeholders in watershed management, and what are some key principles for facilitating constructive dialogue?
2. Can you elaborate on the role of mediation and facilitation in helping stakeholders find common ground and reach agreements in the context of watershed management conflicts?

3. How does conducting conflict mapping and analysis help stakeholders understand the underlying causes and dynamics of conflicts in watershed areas, and what are some common factors driving conflicts in India?
4. What are the benefits of promoting participatory decision-making processes in resolving conflicts over land, water, and natural resources within watershed communities, and how can marginalized groups be empowered to participate effectively?
5. Can you discuss the importance of legal and policy frameworks in providing a basis for resolving disputes and promoting equity and justice in watershed management in India?
6. How do community-based conflict resolution mechanisms contribute to building social cohesion and reducing the need for external intervention in resolving conflicts within watershed areas?
7. What are some key conflict management skills and techniques that stakeholders can develop through capacity building programs, and how do these skills enhance their ability to prevent and resolve conflicts effectively?
8. How can monitoring and evaluation of conflict resolution processes and outcomes inform the refinement of conflict resolution strategies and improve future interventions in watershed management?
9. How do technical interventions such as soil conservation, water harvesting, afforestation, and sustainable agriculture practices contribute to improving watershed health and enhancing the resilience of communities in India?
10. How does the concept of "learning-by-doing" contribute to the adaptive management of watershed projects, and what role does it play in fostering innovation and problem-solving?
11. Can you explain how feedback loops are utilized in adaptive management processes within watershed projects, and how they help in identifying emerging issues and refining project strategies?
12. What are the key components of adaptive experimentation in watershed management, and how can structured experiments contribute to evidence-based decision-making and learning?
13. How do continuous improvement processes contribute to the adaptive capacity of watershed projects, and what mechanisms can be implemented to ensure regular reflection and adjustment of project strategies?

14. Could you elaborate on the role of scenario planning in adaptive management of watershed projects, and how stakeholders can use this approach to anticipate and respond to future uncertainties and risks?
15. What are resilience assessments in the context of watershed management, and how do they help project teams and stakeholders identify strategies for enhancing the resilience of social-ecological systems?
16. How does participatory decision-making enhance the adaptive capacity of watershed projects, and what are some examples of participatory approaches that empower local communities to engage in adaptive management processes?
17. What are some innovative approaches and tools used in watershed management, and how do they address challenges such as complexity, data limitations, and conflicting priorities? Additionally, could you provide examples of lessons learned from implementing these approaches.
18. How can hydrological models like SWAT (Soil and Water Assessment Tool) or MIKE BASIN be effectively utilized to simulate and analyze water flow, sediment transport, and pollutant loading in watershed systems, and what are the key considerations for ensuring accurate and reliable results in watershed planning and management?
19. What are the essential steps involved in developing community scorecards for watershed management projects, and how can these scorecards be used to engage stakeholders in monitoring and evaluating project progress and outcomes?
20. How can predefined criteria and indicators for community scorecards be selected to ensure they accurately reflect the goals and impacts of watershed management projects, and what role do these scorecards play in promoting transparency and accountability among stakeholders?
21. What methods can be used to effectively engage stakeholders in co-creating visions, setting priorities, and co-designing adaptive management strategies for watershed management, and how can these methods enhance the overall success and sustainability of the projects?
22. How can local communities be empowered to take ownership of adaptive management processes, and what capacity-building initiatives are essential to enable them to effectively monitor, analyze, and respond to changes in watershed conditions?

23. What strategies can be implemented to foster a culture of reflection, inquiry, and openness among project teams and stakeholders in watershed management activities, and how can these strategies enhance adaptive learning and innovation?
24. How can project teams and stakeholders be encouraged to view failures as valuable learning opportunities, and what approaches can be used to support experimentation and innovation in developing solutions to complex watershed management challenges?
25. How can integrated watershed management approaches effectively balance environmental conservation, socio-economic development, and community resilience, and what key factors contribute to the successful implementation of such strategies?

6

Remote Sensing in Watershed Management

6.1 Introduction to Remote Sensing in Watershed Management

Remote sensing involves acquiring information about the Earth's surface without physical contact, using satellite or aerial imagery. This technology has revolutionized watershed management by providing detailed and timely data, crucial for effective planning and decision-making.

6.2 Remote Sensing Technologies and Techniques

Satellite Imagery

Satellite imagery involves capturing images of the Earth from space using satellites equipped with sensors. Some of the most notable satellite missions providing high-resolution imagery for large-scale environmental monitoring include Landsat, Sentinel, and MODIS.

a. **Landsat**: Operated by NASA and the US Geological Survey, Landsat satellites have been observing the Earth since 1972. Landsat provides medium-resolution images (15-60 meters per pixel) in both visible and infrared wavelengths, allowing for detailed monitoring of land use, vegetation health, and water bodies.

b. **Sentinel**: Part of the European Space Agency's Copernicus program, the Sentinel satellites (especially Sentinel-2) offer high-resolution (10-20 meters per pixel) optical imagery across multiple spectral bands. They are useful for agricultural monitoring, forest management, and disaster response.

c. **MODIS (Moderate Resolution Imaging Spectroradiometer)**: Installed on NASA's Terra and Aqua satellites, MODIS captures images at moderate resolution (250 meters to 1 kilometer per pixel) but with a wide swath of 2330 kilometers, allowing for frequent global coverage. MODIS data is essential for tracking large-scale environmental changes, such as deforestation, wildfire detection, and sea surface temperature.

Aerial Photography and Drones

Aerial photography and drone technology involve capturing high-resolution images from the air, providing detailed views of specific areas, often at a much higher resolution than satellite imagery.

a. **Aerial Photography**: Taken from manned aircraft, these images can cover large areas with resolutions as fine as a few centimeters per pixel. This technique is particularly useful for creating detailed maps, urban planning, and monitoring changes in land use.

b. **Drones**: Unmanned aerial vehicles (UAVs) equipped with cameras can fly at low altitudes, capturing ultra-high-resolution images (up to sub-centimeter level). Drones are highly versatile and can be deployed quickly to monitor specific areas, inspect infrastructure, and assess disaster impacts. They are invaluable for precision agriculture, forestry, and environmental conservation efforts.

LiDAR (Light Detection and Ranging)

LiDAR technology uses laser pulses to measure distances to the Earth's surface, creating precise 3D models of the terrain.

a. **Elevation Data**: LiDAR can generate highly accurate elevation maps with vertical accuracies of a few centimeters. This data is crucial for topographic mapping, identifying terrain features, and conducting geological surveys.

b. **Flood Modeling**: By providing detailed elevation information, LiDAR helps in creating accurate floodplain maps, predicting flood extents, and developing flood risk management strategies.

c. **Vegetation Analysis**: LiDAR can penetrate forest canopies, providing information about forest structure, biomass, and understory vegetation.

SAR (Synthetic Aperture Radar)

SAR technology uses radar waves to create images of the Earth's surface, capable of penetrating clouds, smoke, and vegetation.

a. **All-Weather Imaging**: SAR can operate in any weather condition, day or night, making it reliable for continuous monitoring. This is particularly useful in regions with frequent cloud cover.

b. **Surface Moisture Monitoring**: SAR can detect changes in soil moisture content, which is important for agriculture, hydrology, and drought assessment.

c. **Land Deformation**: SAR interferometry can measure ground deformation with millimetre accuracy, useful for monitoring earthquakes, landslides, and subsidence.

Multispectral and Hyperspectral Imaging

These imaging techniques capture data across multiple wavelengths beyond the visible spectrum, providing detailed information about the Earth's surface composition.

a. **Multispectral Imaging**: Captures data in a few broad spectral bands (e.g., visible, near-infrared, and shortwave infrared). This data helps in vegetation analysis, land use classification, and water quality assessment. For example, NDVI (Normalized Difference Vegetation Index) is derived from multispectral data to monitor plant health.

b. **Hyperspectral Imaging**: Captures data in hundreds of narrow spectral bands, allowing for detailed identification of materials and their properties. This technique is powerful for mineral exploration, environmental monitoring, and precision agriculture. Hyperspectral imaging can distinguish between different types of vegetation, detect pollutants, and assess soil properties.

Together, these technologies provide comprehensive tools for environmental monitoring, resource management, disaster response, and scientific research, each contributing unique capabilities to understand and manage the Earth's natural resources and hazards.

6.3 Applications of Remote Sensing in Watershed Management

Land Use and Land Cover Mapping

Land use and land cover (LULC) mapping involves classifying the Earth's surface into different categories, such as urban areas, agricultural land, forests, water bodies, and barren land. This is essential for understanding human impacts and natural processes.

a. **Monitoring Changes**: By analyzing satellite imagery and aerial photographs over time, changes in land use and vegetation cover can be tracked. This helps in assessing urban expansion, deforestation, agricultural activities, and natural regeneration.

b. **Human Impact**: LULC maps are crucial for urban planning, infrastructure development, and assessing the environmental impacts of human activities. They help in identifying areas of significant change and implementing sustainable land management practices.

c. **Natural Processes**: These maps also aid in understanding ecological processes, habitat distribution, and biodiversity. They support conservation efforts by highlighting critical habitats and areas requiring protection.

Hydrological Modeling

Hydrological modeling involves simulating the movement, distribution, and quality of water within the hydrological cycle. Remote sensing provides essential inputs for these models.

a. **Precipitation**: Satellites like TRMM and GPM provide global precipitation data, which is critical for modeling rainfall patterns and predicting water availability.

b. **Soil Moisture**: Sensors like SMOS and SMAP measure soil moisture levels, helping to estimate runoff, infiltration, and drought conditions.

c. **Runoff**: Remote sensing data on land cover and topography, combined with precipitation and soil moisture data, help in modeling surface runoff and streamflow, essential for flood forecasting and water resource management.

Soil Erosion and Sedimentation

Soil erosion and sedimentation are significant environmental concerns, affecting land productivity and water quality.

a. **Erosion-Prone Areas**: Remote sensing data can identify areas susceptible to erosion by analyzing vegetation cover, slope, and soil properties. High-resolution imagery helps in detecting gullies and other erosion features.

b. **Sediment Deposition**: By monitoring changes in river courses and deltas, sediment deposition patterns can be mapped. This information is vital for designing soil conservation measures and maintaining healthy aquatic ecosystems.

c. **Conservation Measures**: Implementing soil conservation techniques, such as contour plowing, terracing, and reforestation, can be guided by remote sensing data to reduce erosion and manage sediment loads.

Water Quality Monitoring

Monitoring water quality involves assessing various parameters that indicate the health of water bodies.

a. **Turbidity**: Satellite sensors, such as those on Landsat and Sentinel-2, can measure water clarity by detecting the amount of suspended particles. High turbidity often indicates pollution or sediment runoff.

b. **Chlorophyll Concentration**: Multispectral and hyperspectral sensors can estimate chlorophyll levels, which are indicative of algal blooms and nutrient levels in water bodies. This is crucial for managing eutrophication.

c. **Suspended Sediments**: Remote sensing helps in quantifying suspended sediments, providing insights into erosion, runoff, and sediment transport dynamics. This data is essential for maintaining water quality and aquatic habitats.

Vegetation and Forest Management

Managing vegetation and forests sustainably requires detailed monitoring of various parameters.

a. **Vegetation Health**: Indices like NDVI (Normalized Difference Vegetation Index) derived from satellite data help in assessing plant health, vigor, and stress. This information supports agricultural practices and ecosystem management.

b. **Forest Cover Changes**: Remote sensing tracks deforestation, afforestation, and forest degradation over time. This is critical for implementing policies to protect forests and enhance carbon sequestration.

c. **Biomass Estimation**: Using LiDAR and multispectral data, forest biomass and structure can be estimated, aiding in carbon stock assessments and sustainable forest management practices.

Wetland Mapping and Monitoring

Wetlands are critical ecosystems that require precise mapping and monitoring for conservation.

a. **Detailed Maps**: High-resolution satellite imagery and aerial photographs provide detailed maps of wetland extents and types. These maps help in identifying and protecting wetland habitats.

b. **Monitoring Changes**: Remote sensing tracks changes in wetland areas due to natural and anthropogenic factors, such as climate change, land development, and water management practices.

c. **Conservation Efforts**: Accurate wetland data supports conservation efforts by identifying critical areas for protection, restoration projects, and assessing the impacts of environmental policies.

Flood Monitoring and Management

Effective flood management requires real-time monitoring and accurate flood risk assessment.

a. **Real-Time Monitoring**: Satellites like Sentinel-1 (SAR) provide real-time data on flood extents, even under cloudy conditions. This is crucial for emergency response and disaster management.

b. **Flood Extents**: Remote sensing helps in mapping flood-prone areas, understanding flood dynamics, and predicting future flood scenarios. This information is used to develop flood risk maps and mitigation strategies.

c. **Risk Maps**: Combining topographic data from LiDAR with flood extent data, detailed flood risk maps can be created. These maps guide urban planning, infrastructure development, and community preparedness efforts.

Climate Change Impact Assessment

Assessing the impacts of climate change involves monitoring various environmental parameters over time.

a. **Glaciers and Snow Cover**: Remote sensing tracks changes in glacier extents and snow cover, providing insights into the effects of global warming. This data is crucial for water resource management in regions dependent on glacier melt.

b. **Vegetation**: Monitoring shifts in vegetation types and health helps in understanding the ecological impacts of climate change. Changes in growing seasons, plant distribution, and productivity are key indicators.

c. **Adaptive Strategies**: By studying climate-related changes, adaptive strategies can be developed to mitigate impacts. This includes planning for water scarcity, managing agricultural practices, and protecting vulnerable ecosystems.

These applications of remote sensing technologies enable detailed, large-scale environmental monitoring and support informed decision-making for sustainable resource management and environmental conservation.

6.4 Integration of Remote Sensing with GIS and Other Technologies

Combining remote sensing with Geographic Information Systems (GIS) significantly enhances the capabilities for data analysis, integration, and visualization. This synergy is particularly powerful for applications like watershed management, where diverse datasets need to be analyzed and visualized comprehensively. Here's a detailed description of how GIS enhances remote sensing data for watershed management:

Data Integration

GIS provides a platform for integrating various types of data, both spatial and non-spatial, from multiple sources. This integration is crucial for a holistic understanding of watershed dynamics.

a. **Remote Sensing Data**: Satellite imagery, aerial photographs, LiDAR data, SAR data, and multispectral/hyperspectral images can be imported into GIS. These datasets provide detailed information on land cover, topography, soil moisture, vegetation health, and more.

b. **Ancillary Data**: GIS can incorporate additional datasets such as soil maps, hydrological data, meteorological data, land use records, and demographic information. This helps in creating a more comprehensive analysis framework.

c. **Temporal Analysis**: GIS allows for the integration of time-series data, enabling the monitoring of changes over time. This is vital for tracking trends in land use, vegetation cover, water quality, and climate impacts.

Spatial Analysis

GIS tools offer robust spatial analysis capabilities, which are essential for understanding the complex interactions within a watershed.

a. **Overlay Analysis**: By overlaying different data layers, GIS can help identify relationships and patterns. For example, combining elevation data with land use and precipitation data can help in identifying areas at risk of erosion or flooding.

b. **Buffer Analysis**: GIS can create buffer zones around features like rivers, lakes, and wetlands to analyze the impact of surrounding land use activities. This is useful for managing riparian zones and protecting water quality.

c. **Network Analysis**: Using hydrological data, GIS can model river networks, watersheds, and drainage basins. This helps in understanding the flow of water, sediment transport, and identifying critical points for intervention.

d. **Statistical Analysis**: GIS supports statistical analysis of spatial data, allowing for the quantification of changes and the identification of significant trends and anomalies.

Visualization

One of the key strengths of GIS is its ability to create detailed, interactive maps and visualizations that make complex data more understandable and actionable.

a. **Thematic Maps**: GIS can generate thematic maps that highlight specific attributes such as land cover types, soil erosion risk, flood zones, and vegetation health. These maps are valuable for decision-making and communication.

b. **3D Visualization**: Integrating LiDAR data with GIS allows for the creation of 3D terrain models. These models provide a realistic representation of the watershed's topography, which is essential for flood modeling and infrastructure planning.

c. **Dynamic Visualization**: GIS can produce animations and time-lapse visualizations to show changes over time. This is particularly useful for presenting the impacts of climate change, land use changes, and other temporal dynamics.

d. **Interactive Dashboards**: GIS platforms can create interactive dashboards that allow users to explore data layers, run analyses, and visualize results dynamically. This enhances stakeholder engagement and supports informed decision-making.

Applications in Watershed Management

The integration of remote sensing with GIS offers a powerful toolkit for comprehensive watershed management. It enhances the ability to analyze complex spatial data, visualize dynamic changes, and make informed decisions for sustainable resource management and environmental conservation. This combination is instrumental in addressing the challenges of water resource management, ecological conservation, and climate change adaptation. Combining remote sensing with GIS enables a wide range of applications in watershed management:

a. **Land Use and Land Cover Mapping**: GIS integrates remote sensing data to create detailed LULC maps, which are essential for planning and managing land resources within the watershed.

b. **Hydrological Modeling**: GIS supports the integration of precipitation, soil moisture, and runoff data from remote sensing into hydrological models. This helps in predicting water availability, flood risks, and drought impacts.

c. **Soil Erosion and Sedimentation**: GIS tools can analyze topographic data from LiDAR and other sources to identify erosion-prone areas and sediment deposition zones. This information guides soil conservation efforts.

d. **Water Quality Monitoring**: By integrating remote sensing data on turbidity, chlorophyll concentration, and suspended sediments, GIS helps in monitoring and managing water quality within the watershed.

e. **Vegetation and Forest Management**: GIS facilitates the tracking of vegetation health and forest cover changes using remote sensing data. This supports sustainable forest management practices and biodiversity conservation.

f. **Wetland Mapping and Monitoring**: GIS integrates high-resolution imagery and other remote sensing data to map and monitor wetland areas. This aids in the conservation and restoration of these critical habitats.

g. **Flood Monitoring and Management**: Combining real-time SAR data with GIS, flood extents can be mapped accurately. GIS-based flood risk maps support disaster preparedness and response efforts.

h. **Climate Change Impact Assessment**: GIS helps in visualizing and analyzing the impacts of climate change on glaciers, snow cover, vegetation, and other watershed components. This information is crucial for developing adaptive strategies.

Table: This table provides an overview of how each technology is applied in watershed management, along with brief descriptions and potential Indian examples for context.

S. No.	Technology	Technology	Technology
1	Remote Sensing	Land Use and Land Cover Mapping	Monitoring changes in land use and vegetation cover to understand human impacts and natural processes.
2	Remote Sensing	Hydrological Modeling	Deriving inputs like precipitation, soil moisture, and runoff for hydrological models to predict water availability and flood risks.
3	Remote Sensing	Soil Erosion and Sedimentation Mapping	Identifying erosion-prone areas and sediment deposition to implement soil conservation measures.
4	Remote Sensing	Water Quality Monitoring	Assessing parameters like turbidity, chlorophyll concentration, and suspended sediments to monitor water health.
5	Remote Sensing	Vegetation and Forest Management	Tracking vegetation health, forest cover changes, and biomass for sustainable forest management.
6	Remote Sensing	Wetland Mapping and Monitoring	Providing detailed maps of wetland areas and monitoring changes to support conservation efforts.
7	Remote Sensing	Flood Monitoring and Management	Real-time monitoring of flood events, assessing flood extents, and developing risk maps.
8	Remote Sensing	Climate Change Impact Assessment	Studying impacts on glaciers, snow cover, and vegetation to develop adaptive strategies.
9	Geographic Information Systems (GIS)	Spatial Analysis	Integrating various data layers, performing spatial analysis, and creating comprehensive maps for watershed management.
10	Geographic Information Systems (GIS)	Decision Support Systems (DSS)	Developing decision support systems to assist in planning and management activities.
11	Geographic Information Systems (GIS)	Watershed Modeling	Building hydrological and ecological models to simulate watershed processes and assess management scenarios.
12	Geographic Information Systems (GIS)	Risk Mapping and Assessment	Identifying and assessing risks related to flooding, erosion, water quality, and land use changes.

S. No.	Technology	Technology	Technology
13	Geographic Information Systems (GIS)	Land Use Planning and Zoning	Guiding land use planning and zoning regulations to minimize environmental impact and promote sustainable development.
14	LiDAR (Light Detection and Ranging)	Terrain Analysis	Providing precise elevation data for topographic mapping and flood modeling.
15	LiDAR (Light Detection and Ranging)	Vegetation Structure Mapping	Assessing vegetation structure and biomass for ecosystem monitoring and management.
16	LiDAR (Light Detection and Ranging)	Floodplain Mapping	Mapping floodplains and flood hazard zones for flood risk assessment and management.
17	LiDAR (Light Detection and Ranging)	Urban Growth Monitoring	Monitoring urban expansion and land use changes to inform urban planning decisions.
18	LiDAR (Light Detection and Ranging)	Infrastructure Planning	Assessing the impact of infrastructure projects on watersheds and optimizing planning decisions.
19	IoT (Internet of Things)	Real-time Monitoring	Integrating remote sensing data with ground-based sensors for continuous monitoring of environmental parameters.
20	IoT (Internet of Things)	Water Quality Sensors	Deploying sensors to monitor water quality parameters such as pH, dissolved oxygen, and temperature.
21	IoT (Internet of Things)	Weather Stations	Installing weather stations to collect meteorological data for hydrological modelling and forecasting.
22	IoT (Internet of Things)	Soil Moisture Sensors	Placing sensors to measure soil moisture content for agricultural water management and erosion control.
23	Artificial Intelligence (AI)	Image Classification	Using machine learning algorithms to classify land cover types and vegetation species from satellite imagery.
24	Artificial Intelligence (AI)	Flood Prediction	Developing AI models to predict floods based on historical data, weather forecasts, and hydrological parameters.
25	Artificial Intelligence (AI)	Crop Health Monitoring	Analyzing satellite imagery to assess crop health, detect diseases, and optimize agricultural practices.

S. No.	Technology	Technology	Technology
26	Artificial Intelligence (AI)	Pattern Recognition	Identifying patterns of land use change, vegetation dynamics, and urban growth from satellite images.
27	Artificial Intelligence (AI)	Risk Assessment	Integrating multiple data sources and AI algorithms to assess risks related to floods, erosion, and water pollution.
28	IoT (Internet of Things)	Real-time Monitoring	Integrating remote sensing data with ground-based sensors for continuous monitoring of environmental parameters.
29	IoT (Internet of Things)	Water Quality Sensors	Deploying sensors to monitor water quality parameters such as pH, dissolved oxygen, and temperature.
30	IoT (Internet of Things)	Weather Stations	Installing weather stations to collect meteorological data for hydrological modeling and forecasting.
31	IoT (Internet of Things)	Soil Moisture Sensors	Placing sensors to measure soil moisture content for agricultural water management and erosion control.
32	Artificial Intelligence (AI)	Image Classification	Using machine learning algorithms to classify land cover types and vegetation species from satellite imagery.
33	Artificial Intelligence (AI)	Flood Prediction	Developing AI models to predict floods based on historical data, weather forecasts, and hydrological parameters.
34	Artificial Intelligence (AI)	Crop Health Monitoring	Analyzing satellite imagery to assess crop health, detect diseases, and optimize agricultural practices.
35	Artificial Intelligence (AI)	Pattern Recognition	Identifying patterns of land use change, vegetation dynamics, and urban growth from satellite images.
36	Artificial Intelligence (AI)	Risk Assessment	Integrating multiple data sources and AI algorithms to assess risks related to floods, erosion, and water pollution.

6.5 Challenges and Limitations of Remote Sensing in Watershed Management

Data Resolution and Accuracy

Ensuring high resolution and accurate data is critical for the effectiveness of remote sensing applications in watershed management.

a. **Resolution**: The resolution of remote sensing data refers to the smallest detail that can be detected. High-resolution data provides more detailed information, which is essential for precise analysis. For instance, in land use and land cover mapping, high-resolution imagery (e.g., 1 meter per pixel) can differentiate between different types of vegetation and small-scale land use changes. Similarly, in hydrological modeling, high-resolution digital elevation models (DEMs) from LiDAR data are crucial for accurate terrain analysis and flood prediction.

b. **Accuracy**: Accuracy pertains to the correctness of the data in representing the real-world scenario. This includes both positional accuracy (how closely the data's geographic position matches the true location) and thematic accuracy (how accurately the data represents the classified themes, such as land cover types). High accuracy is essential for reliable analysis and decision-making. For example, inaccurate elevation data could lead to erroneous flood risk assessments.

Cost and Accessibility

Balancing the cost of obtaining high-resolution imagery with its accessibility is a significant consideration in remote sensing applications.

a. **Cost**: High-resolution satellite imagery and advanced data sources like LiDAR and hyperspectral sensors can be expensive. Costs can include the acquisition of data, the technology needed to process and analyze the data, and the licensing fees. For many stakeholders, especially in developing regions, these costs can be prohibitive.

b. **Accessibility**: Ensuring that the data is accessible to all relevant stakeholders, including government agencies, researchers, and local communities, is crucial. Open-access data from missions like Landsat and Sentinel has greatly enhanced accessibility. However, there remains a gap in access to high-resolution commercial imagery and advanced data products, which can limit the ability of some stakeholders to utilize these resources fully.

Technical Expertise

The use of remote sensing data requires a high level of technical expertise to process, analyze, and interpret the data effectively.

a. **Data Processing**: Processing raw remote sensing data involves several steps, including calibration, correction (e.g., atmospheric and geometric corrections), and transformation (e.g., creating indices like NDVI). This requires knowledge of specialized software and techniques.

b. **Data Analysis**: Analyzing remote sensing data to extract meaningful information involves various methods, such as image classification, change detection, and spatial analysis within a GIS. Expertise in these methods is essential to derive accurate and actionable insights.

c. **Interdisciplinary Knowledge**: Effective use of remote sensing data in watershed management often requires interdisciplinary knowledge. For example, understanding hydrological processes, ecological principles, and geospatial technologies is necessary to integrate and analyze data comprehensively.

Temporal Frequency

Ensuring frequent data acquisition is essential to monitor dynamic changes within watersheds and respond to them effectively.

a. **Dynamic Monitoring**: Watersheds are dynamic systems that can change rapidly due to factors such as precipitation, land use changes, and human activities. Frequent data acquisition allows for timely monitoring of these changes. For instance, during the rainy season, frequent monitoring can help track flood events in real-time.

b. **Temporal Analysis**: High temporal frequency is crucial for time-series analysis, which is needed to understand trends and patterns over time. For example, monitoring vegetation health throughout the growing season or tracking changes in water quality over months and years requires frequent data updates.

c. **Data Sources**: Satellites like MODIS offer daily to near-daily observations, which are valuable for continuous monitoring, albeit at a coarser resolution. Conversely, high-resolution satellites may revisit the same area less frequently. Combining data from multiple sources can help achieve a balance between resolution and temporal frequency.

Combining remote sensing with GIS for watershed management involves addressing several critical factors:

a. **Data Resolution and Accuracy**: High-resolution and accurate data is essential for detailed and reliable analysis.

b. **Cost and Accessibility**: Balancing the cost of high-resolution data with its accessibility to all stakeholders is crucial for widespread adoption and effective management.

c. **Technical Expertise**: Skilled personnel are needed to process, analyze, and interpret remote sensing data accurately.

d. **Temporal Frequency**: Frequent data acquisition is necessary to monitor and respond to dynamic changes in watersheds effectively.

By addressing these challenges, remote sensing and GIS can significantly enhance the capacity for informed decision-making and sustainable management of watershed resources.

6.6 Future Trends and Innovations in Remote Sensing for Watershed Management

Advancements in Satellite Technology

Recent advancements in satellite technology have significantly enhanced the capabilities of remote sensing for environmental monitoring and management.

a. **Improved Resolution**: Modern satellites are capable of capturing imagery at much higher resolutions than ever before. For example, commercial satellites like WorldView-3 can provide imagery with a resolution of 31 centimeters per pixel, allowing for detailed analysis of small-scale features such as individual trees or infrastructure components. Higher resolution improves the accuracy and detail of land cover maps, urban planning, and environmental monitoring.

b. **Increased Frequency**: New satellite constellations, such as those launched by Planet Labs, provide daily global coverage at high resolution. This increased temporal frequency allows for near-real-time monitoring of dynamic environmental changes, such as deforestation, crop growth, and disaster events.

c. **Advanced Sensors**: Innovations in sensor technology have expanded the range of data that can be collected from space. Hyperspectral sensors capture data across hundreds of narrow spectral bands, enabling detailed analysis of material composition and health. Synthetic Aperture Radar (SAR) sensors, which operate independently of weather and light conditions, provide critical data for monitoring surface deformation, moisture content, and structural integrity.

Artificial Intelligence and Machine Learning

The integration of artificial intelligence (AI) and machine learning (ML) techniques has revolutionized the processing and analysis of remote sensing data.

a. **Automated Data Processing**: AI and ML algorithms can automatically process large volumes of remote sensing data, performing tasks such as image classification, object detection, and change detection more quickly and accurately than traditional methods. For instance, convolutional neural networks (CNNs) can classify land cover types with high precision.

b. **Predictive Modeling**: Machine learning models can analyze historical and real-time data to predict future trends and events. For example, ML algorithms can forecast crop yields based on multispectral imagery, weather data, and soil moisture levels, helping in agricultural planning and food security.

c. **Anomaly Detection**: AI systems can identify anomalies or unexpected changes in environmental data, such as sudden deforestation, illegal mining activities, or algal blooms, enabling rapid response and mitigation efforts.

Integration with Internet of Things (IoT)

Combining remote sensing data with ground-based IoT sensors enhances the capability for real-time environmental monitoring and decision-making.

a. **Real-Time Monitoring**: IoT sensors can provide continuous, real-time data on various environmental parameters such as soil moisture, air and water quality, temperature, and precipitation. This data can be integrated with satellite observations to create a comprehensive and up-to-date picture of the watershed's health and dynamics.

b. **Enhanced Data Accuracy**: Ground-based sensors validate and complement remote sensing data, improving the accuracy and reliability of environmental models. For example, soil moisture sensors can provide ground truth data that calibrates and refines satellite-derived moisture estimates.

c. **Decision Support Systems**: IoT-enabled systems can be integrated into decision support platforms, providing stakeholders with actionable insights. For example, farmers can receive real-time alerts about irrigation needs based on combined data from satellites and soil moisture sensors.

Crowdsourcing and Citizen Science

Engaging communities in data collection and validation efforts through crowdsourcing and citizen science enhances the breadth and depth of environmental monitoring.

a. **Data Collection**: Citizen scientists can contribute valuable data by reporting local observations, such as wildlife sightings, plant phenology, pollution levels, and weather conditions. Mobile apps and online platforms facilitate the collection and submission of these data points.

b. **Data Validation**: Crowdsourced data can help validate and improve the accuracy of remote sensing products. For instance, local observations of land use changes or water quality can be used to verify satellite-derived classifications and measurements.

c. **Community Engagement**: Involving the public in scientific research fosters greater awareness and stewardship of local environments. Community participation in data collection can lead to more informed and engaged citizens, promoting sustainable practices and conservation efforts.

The combination of advancements in satellite technology, artificial intelligence, integration with IoT, and crowdsourcing through citizen science represents a significant leap forward in environmental monitoring and management:

a. **Advancements in Satellite Technology**: Higher resolution, increased frequency, and advanced sensors provide more detailed and timely data for environmental analysis.

b. **Artificial Intelligence and Machine Learning**: AI and ML enhance data processing capabilities and predictive modeling, allowing for more accurate and efficient analysis.

c. **Integration with IoT**: IoT sensors complement remote sensing data, enabling real-time monitoring and improving the accuracy of environmental assessments.

d. **Crowdsourcing and Citizen Science**: Engaging the public in data collection and validation efforts broadens the scope of monitoring and fosters community involvement in environmental stewardship.

These innovations collectively enhance our ability to monitor, understand, and manage the environment, contributing to more effective and sustainable resource management practices.

6.7 Case Studies of Remote Sensing in Watershed Management

Case Study 1: Land Cover Change Detection in the Amazon Basin using Landsat Imagery

Background: The Amazon Basin is a critical region for global biodiversity and climate regulation. However, it faces significant deforestation pressures due to agriculture, logging, and urbanization.

Objective: To monitor and analyze land cover changes over time to understand the extent and impact of deforestation.

Methodology

- **Data Collection**: Landsat imagery from multiple time periods was used.
- **Analysis**: Change detection techniques, such as Normalized Difference Vegetation Index (NDVI), were employed to identify areas of deforestation and reforestation.
- **Validation**: Ground truthing and comparison with historical land use data ensured accuracy.

Findings

- Significant deforestation was detected, especially in areas close to roads and rivers.
- Patterns of land use change indicated a shift from forest to agricultural lands.
- Some areas showed signs of natural reforestation and forest recovery.

Impact

- The study provided actionable data for policymakers to design conservation strategies and enforce regulations against illegal logging.
- It also highlighted the importance of monitoring and protecting forested areas to maintain biodiversity and carbon sequestration capacities.

Case Study 2: Land Cover Change Detection in the Western Ghats using Landsat Imagery

Background

The Western Ghats, a UNESCO World Heritage site, is a biodiversity hotspot in India known for its rich flora and fauna. However, this region faces significant environmental pressures due to agricultural expansion, plantation forestry, urbanization, and infrastructure development. Monitoring land cover changes in the Western Ghats is crucial for understanding the impacts of human activities and for devising conservation strategies.

Objective

To monitor and analyze land cover changes in the Western Ghats over time to understand the extent and impact of deforestation and land use changes.

Methodology

Data Collection

- Satellite Imagery: Landsat imagery from different time periods (e.g., 1990, 2000, 2010, and 2020) was utilized to capture changes over the decades.

- **Ancillary Data**: Historical land use data, topographic maps, and local climate data were used to supplement the analysis.

Analysis

- **Preprocessing**: The Landsat images underwent preprocessing steps including atmospheric correction, geometric correction, and radiometric normalization to ensure data quality and comparability.
- **Change Detection Techniques**: Normalized Difference Vegetation Index (NDVI) and other indices like Enhanced Vegetation Index (EVI) were used to detect changes in vegetation cover. These indices help in identifying deforested areas and areas undergoing reforestation.
- **Classification**: Supervised classification methods, such as Support Vector Machines (SVM) and Random Forests (RF), were used to categorize the land cover into classes such as forest, agriculture, urban, water bodies, and barren land.
- **Temporal Analysis**: Time-series analysis was performed to identify trends and patterns of land cover change over the selected periods.

Validation

- **Ground Truthing**: Field visits and GPS data collection were conducted to validate the classification results. Local knowledge from community members and experts was also incorporated.
- **Comparison with Historical Data**: The results were compared with historical land use records to ensure accuracy and consistency.

Findings

- **Deforestation Trends**: Significant deforestation was detected, particularly in areas adjacent to expanding agricultural lands and plantation forestry regions. Deforestation hotspots were identified in districts like Wayanad and Kodagu.
- **Urbanization**: Rapid urban expansion was observed around cities like Pune and Coimbatore, leading to the conversion of forested areas into urban settlements.
- **Agricultural Expansion**: Large tracts of forest land were converted to agricultural lands, primarily for cash crops such as coffee, tea, and spices.
- **Reforestation**: Some regions, particularly those under conservation projects and community-managed forests, showed signs of natural reforestation and forest recovery. These areas were mostly located in protected regions and buffer zones.

- **Impact of Infrastructure**: Construction of new roads and highways contributed to deforestation and fragmentation of habitats, making certain areas more accessible and prone to land use changes.

Impact

- **Policy Implications**: The study provided critical data to policymakers for designing effective conservation strategies. It emphasized the need for stricter enforcement of regulations against illegal deforestation and land conversion.
- **Conservation Efforts**: The findings supported the expansion of protected areas and the implementation of sustainable land management practices. Community-based forest management and reforestation projects were highlighted as successful models.
- **Urban Planning**: The data helped in guiding urban planning and development policies to minimize environmental impact. Zoning regulations and green belt designs were recommended to control urban sprawl.
- **Awareness and Education**: The study raised awareness among local communities and stakeholders about the importance of preserving the Western Ghats. Educational campaigns were launched to promote sustainable practices and conservation efforts.
- **Biodiversity and Climate Regulation**: By highlighting the importance of the Western Ghats in biodiversity conservation and carbon sequestration, the study underscored the need to protect these forests to mitigate climate change and preserve ecosystem services.

Conclusion

The case study of land cover change detection in the Western Ghats using Landsat imagery demonstrates the power of remote sensing and GIS in monitoring environmental changes. The comprehensive analysis provided actionable insights for policymakers, conservationists, and local communities, leading to informed decisions and sustainable management of this vital ecological region. Through continued monitoring and the integration of advanced technologies like AI, ML, and IoT, the efforts to protect and conserve the Western Ghats can be further strengthened.

Case Study 3: Flood Risk Assessment and Management in the Mississippi River Basin using SAR Data

Background: The Mississippi River Basin frequently experiences severe flooding, causing extensive economic damage and loss of life.

Objective: To develop a comprehensive flood risk assessment model and improve flood management strategies.

Methodology

- **Data Collection**: Synthetic Aperture Radar (SAR) data from satellites like Sentinel-1 was used due to its ability to penetrate clouds and vegetation.
- **Analysis**: SAR imagery was processed to map flood extents and water levels during various flood events.
- **Integration**: Data was integrated with hydrological models and GIS to create detailed flood risk maps.

Findings

- Accurate flood extent maps were produced for recent flood events, helping identify high-risk areas.
- The integration of SAR data with hydrological models improved the accuracy of flood predictions.

Impact

- The study enhanced flood preparedness by providing detailed risk maps to emergency management agencies.
- It supported the development of better flood mitigation infrastructures, such as levees and drainage systems.

Case Study 4: Flood Risk Assessment and Management in the Godavari River Basin using SAR Data

Background

The Godavari River Basin, one of the largest river basins in India, frequently experiences severe flooding, particularly during the monsoon season. These floods have significant socio-economic impacts, affecting millions of people, destroying agricultural lands, and damaging infrastructure. Effective flood risk assessment and management are crucial for mitigating these impacts and enhancing community resilience.

Objective

To develop a comprehensive flood risk assessment model and enhance flood management strategies in the Godavari River Basin using Synthetic Aperture Radar (SAR) data.

Methodology

Data Collection

- **SAR Data**: Data from Sentinel-1 satellites, equipped with C-band SAR sensors, were used. Sentinel-1's ability to penetrate cloud cover and vegetation makes it ideal for flood mapping in regions with frequent cloud cover.
- **Historical Data**: Past flood records, topographic maps, hydrological data, and land use/land cover maps were collected to provide context and support the analysis.

Analysis

- **Preprocessing**: SAR images were pre-processed to remove noise, correct for geometric distortions, and calibrate the data. This included speckle filtering, radiometric calibration, and terrain correction.
- **Flood Mapping**: The processed SAR data was used to identify flood extents. Techniques such as change detection, flood index mapping, and thresholding were applied to distinguish between flooded and non-flooded areas.
- **Hydrological Modeling**: Data from SAR imagery was integrated with hydrological models, such as HEC-HMS (Hydrologic Modeling System) and HEC-RAS (River Analysis System), to simulate river flow and water levels. This integration helped in refining flood prediction models.
- **GIS Integration**: GIS tools were used to overlay flood extent maps with spatial data on population density, infrastructure, and critical facilities. This spatial integration facilitated the creation of detailed flood risk maps.

Validation

- **Ground Truthing**: Field surveys and data from local weather stations were used to validate the SAR-derived flood maps. Comparison with satellite imagery from optical sensors and historical flood records ensured the accuracy of the flood extents.
- **Model Calibration**: Hydrological models were calibrated and validated using observed river discharge data and water level measurements from gauge stations across the basin.

Findings

- **Accurate Flood Extent Maps**: The study successfully produced precise flood extent maps for recent flood events, including the 2019 and 2020 floods. These maps identified high-risk areas along the riverbanks and low-lying regions prone to inundation.
- **Improved Flood Predictions**: The integration of SAR data with hydrological models significantly enhanced the accuracy of flood predictions. This integration enabled more precise forecasting of flood peaks and water levels, reducing uncertainties in flood modeling.
- **Spatial Distribution of Risk**: Analysis revealed that areas near major tributaries, such as the Pranhita and Penganga rivers, were particularly vulnerable to flooding. Urban areas, agricultural zones, and infrastructure along these tributaries were identified as high-risk zones.
- **Impact of Land Use Changes**: The study highlighted the impact of land use changes, such as deforestation and urbanization, on flood dynamics. Areas with reduced vegetation cover and increased impervious surfaces showed higher flood susceptibility.

Impact

- **Enhanced Flood Preparedness**: The flood risk maps generated from the study have been instrumental for local and state emergency management agencies. These maps provide critical information for flood forecasting, evacuation planning, and disaster response strategies.
- **Improved Infrastructure Planning**: The findings supported the design and implementation of better flood mitigation infrastructures, such as levees, embankments, and improved drainage systems. Areas identified as high-risk were prioritized for infrastructure upgrades to enhance flood resilience.
- **Policy Development and Planning**: The study informed policy development and urban planning initiatives, promoting the adoption of flood-resistant building codes and land use regulations. It also supported the development of community-based flood management plans and early warning systems.
- **Community Engagement and Awareness**: The project engaged local communities through workshops, training sessions, and awareness campaigns, enhancing their understanding of flood risks and promoting participatory flood management practices. This grassroots engagement fostered a culture of preparedness and resilience among the local populations.

Conclusion

The application of SAR data for flood risk assessment in the Godavari River Basin has demonstrated the effectiveness of remote sensing and GIS technologies in enhancing flood management strategies. By providing accurate flood maps and improving hydrological modeling, the study has significantly contributed to better flood preparedness, infrastructure planning, and community resilience. Continued monitoring and the integration of advanced technologies such as AI, ML, and IoT will further strengthen flood risk management efforts in the region, ensuring sustainable development and safeguarding lives and livelihoods against future flood events.

Case Study 5: Monitoring Water Quality in the Great Lakes using Multispectral Satellite Data

Background: The Great Lakes are a vital freshwater resource but face pollution from agricultural runoff, industrial discharge, and urban development.

Objective: To monitor water quality parameters and identify pollution sources to protect this vital resource.

Methodology

- **Data Collection**: Multispectral satellite data from sensors like MODIS and Sentinel-2 were utilized.
- **Analysis**: Algorithms were applied to estimate water quality indicators such as chlorophyll concentration, turbidity, and suspended sediment load.
- **Validation**: Data was validated using in-situ water quality measurements from monitoring stations.

Findings

- Areas with high nutrient concentrations and algal blooms were identified, indicating potential sources of pollution.
- Seasonal variations in water quality parameters were observed, correlating with agricultural practices and urban runoff.

Impact

- The study provided critical information for water resource managers to implement pollution control measures.
- It raised awareness about the impact of land use on water quality, promoting sustainable practices in agriculture and urban planning.

Case Study 6: Monitoring Water Quality in the Ganges River using Multispectral Satellite Data

Background

The Ganges River, one of the most important and sacred rivers in India, is a crucial freshwater resource for millions of people. However, the river faces significant pollution from agricultural runoff, industrial discharge, and urban development. Monitoring water quality is essential for protecting the health of the river and the communities that depend on it.

Objective

To monitor water quality parameters and identify pollution sources in the Ganges River using multispectral satellite data, thereby aiding in the protection and management of this vital resource.

Methodology

Data Collection

- **Multispectral Satellite Data**: Data from sensors such as MODIS (Moderate Resolution Imaging Spectroradiometer) on the Terra and Aqua satellites, and Sentinel-2 (part of the Copernicus Programme) were used. These sensors provide data in multiple spectral bands, allowing for the analysis of various water quality parameters.
- ***In-situ* Measurements**: Water quality data from monitoring stations along the Ganges River were collected to validate the satellite-derived estimates.

Analysis

- **Preprocessing**: The satellite data underwent preprocessing steps including atmospheric correction, geometric correction, and radiometric calibration to ensure accuracy and comparability.
- **Algorithm Application**: Specific algorithms were applied to the multispectral data to estimate water quality indicators:
- **Chlorophyll Concentration**: Algorithms like the Normalized Difference Chlorophyll Index (NDCI) were used to estimate the concentration of chlorophyll-a, an indicator of algal biomass.
- **Turbidity**: Reflectance ratios in certain spectral bands were used to estimate turbidity levels, which indicate the presence of suspended particles in the water.
- **Suspended Sediment Load**: Algorithms were applied to estimate the concentration of suspended sediments, which affect water clarity and quality.

- **Temporal Analysis**: Time-series analysis was conducted to observe seasonal and temporal variations in water quality parameters, correlating these with potential pollution sources such as agricultural practices and industrial activities.

Validation

- **Ground Truthing**: The satellite-derived estimates were validated using in-situ measurements from water quality monitoring stations along the river. Parameters such as chlorophyll concentration, turbidity, and sediment load were compared to ensure accuracy.
- **Comparison with Historical Data**: The results were compared with historical water quality data to understand trends and validate the findings.

Findings

- **Pollution Hotspots**: The study identified areas with high nutrient concentrations and frequent algal blooms, particularly in regions downstream of major cities like Kanpur, Varanasi, and Patna. These hotspots indicated significant pollution from urban runoff and industrial discharges.
- **Agricultural Runoff**: High levels of turbidity and suspended sediments were observed in agricultural regions, especially during the monsoon season. The increase in turbidity was correlated with agricultural runoff and soil erosion.
- **Seasonal Variations**: The analysis revealed distinct seasonal variations in water quality parameters. For instance, higher chlorophyll concentrations and turbidity levels were observed during the monsoon season, correlating with increased runoff and agricultural activity.
- **Impact of Industrial Discharge**: Industrial areas along the river were associated with elevated levels of pollutants, particularly in stretches near major industrial hubs.

Impact

- **Water Resource Management**: The study provided critical information for water resource managers and policymakers to implement targeted pollution control measures. For example, areas identified as pollution hotspots can be prioritized for intervention and cleanup efforts.
- **Sustainable Practices**: The findings raised awareness about the impact of land use on water quality, promoting sustainable agricultural and

industrial practices. Farmers were encouraged to adopt practices that reduce runoff, such as contour plowing and the use of buffer strips.

- **Regulatory Measures**: The data supported the enforcement of stricter regulations on industrial discharge and urban waste management. Policies aimed at reducing the release of untreated sewage and industrial effluents into the river were strengthened.
- **Public Awareness and Community Involvement**: The study engaged local communities and raised public awareness about the importance of maintaining water quality. Educational campaigns and community involvement in river monitoring activities were promoted to foster a sense of stewardship and collective responsibility.
- **Health and Ecosystem Protection**: By identifying pollution sources and trends, the study helped protect public health and the river's ecosystem. Efforts to reduce pollution directly contributed to improving the quality of drinking water, reducing the incidence of waterborne diseases, and preserving aquatic biodiversity.

Conclusion

The case study on monitoring water quality in the Ganges River using multispectral satellite data demonstrates the effectiveness of remote sensing in managing and protecting vital freshwater resources. The comprehensive analysis provided actionable insights for improving water quality management, promoting sustainable practices, and enhancing public awareness. Continued monitoring and the integration of advanced technologies will further strengthen efforts to preserve the health and vitality of the Ganges River, ensuring it remains a lifeline for millions of people.

Case Study 7: Soil Erosion Mapping in the Loess Plateau of China using LiDAR Technology

Background: The Loess Plateau in China is highly susceptible to soil erosion due to its loose, fine-grained soil and intensive agricultural activities.

Objective: To map soil erosion patterns and assess the effectiveness of erosion control measures.

Methodology

- **Data Collection**: High-resolution LiDAR data was collected to capture detailed topographic information.
- **Analysis**: Digital Elevation Models (DEMs) were created from LiDAR data to identify erosion features such as gullies and rills.

- **Integration**: LiDAR data was combined with soil type and land use data to assess erosion risk.

Findings

- The study identified key erosion-prone areas, including steep slopes and regions with minimal vegetation cover.
- It showed that areas with implemented erosion control measures, like terracing and reforestation, had significantly reduced erosion rates.

Impact

- The findings supported the design and implementation of targeted soil conservation practices.
- The study demonstrated the effectiveness of existing erosion control measures, encouraging their wider adoption.

Case Study 8: Soil Erosion Mapping in the Chambal River Basin using LiDAR Technology

Background

The Chambal River Basin, part of the greater Yamuna River Basin in India, is known for its severe soil erosion problems, particularly in the Chambal ravines. The region's fragile soil structure and intensive agricultural activities exacerbate soil erosion, leading to significant land degradation and loss of agricultural productivity.

Objective

To map soil erosion patterns and assess the effectiveness of erosion control measures in the Chambal River Basin using LiDAR technology.

Methodology

Data Collection:

- **LiDAR Data**: High-resolution LiDAR (Light Detection and Ranging) data was collected using airborne LiDAR sensors. This technology provided detailed topographic information essential for identifying micro-topographic features.
- **Supplementary Data**: Additional data on soil types, land use, rainfall patterns, and historical erosion rates were collected to complement the LiDAR data.

Analysis

- **Digital Elevation Models (DEMs)**: High-resolution DEMs were created from the LiDAR data. These models accurately represented the terrain and were used to identify erosion features such as gullies, rills, and sediment deposits.

- **Erosion Feature Identification**: Algorithms were applied to the DEMs to detect and map erosion features. This involved analyzing slope gradients, curvature, and other topographic indices.
- **Erosion Risk Assessment**: The LiDAR-derived topographic data was combined with soil type and land use data to assess erosion risk. Factors such as soil erodibility, land cover, and land management practices were considered in the assessment.
- **Effectiveness of Control Measures**: Areas with implemented erosion control measures, such as terracing, check dams, and reforestation, were analyzed to assess their impact on reducing erosion rates.

Integration

- **GIS Integration**: The LiDAR data and supplementary datasets were integrated into a Geographic Information System (GIS) for spatial analysis and visualization. This allowed for the creation of comprehensive erosion risk maps and the evaluation of control measures.

Validation

- **Ground Truthing**: Field surveys and observations were conducted to validate the LiDAR-derived erosion features and risk assessments. Soil erosion measurements and comparisons with historical data ensured the accuracy of the findings.

Findings

- **Erosion-Prone Areas**: The study identified key erosion-prone areas, including steep slopes, bare lands, and regions with minimal vegetation cover. The Chambal ravines were highlighted as critical erosion hotspots.
- **Impact of Control Measures**: The analysis showed that areas with implemented erosion control measures, such as terracing and reforestation, had significantly reduced erosion rates. Terraced fields and reforested areas exhibited improved soil stability and reduced sediment runoff.
- **Spatial Patterns**: Erosion patterns were closely linked to land use practices and topographic features. Agricultural lands on steep slopes and areas with poor vegetation cover were more prone to severe erosion.

Impact

- **Targeted Soil Conservation Practices**: The findings supported the design and implementation of targeted soil conservation practices. Specific recommendations included expanding terracing in hilly agricultural areas, promoting reforestation in degraded lands, and constructing additional check dams in erosion-prone regions.
- **Policy and Planning**: The study provided valuable data for policymakers and land managers to develop effective soil conservation strategies. It emphasized the need for integrated land management approaches that combine engineering solutions with sustainable agricultural practices.
- **Community Awareness and Participation**: The project engaged local communities and raised awareness about the importance of soil conservation. Training programs and workshops were conducted to educate farmers on erosion control techniques and sustainable land use practices.
- **Validation of Existing Measures**: The study validated the effectiveness of existing erosion control measures, encouraging their wider adoption. Successful case studies of terracing and reforestation were shared with neighboring regions to promote best practices.

The case study on soil erosion mapping in the Chambal River Basin using LiDAR technology demonstrates the powerful application of high-resolution remote sensing in environmental management. By providing detailed topographic data and accurate erosion risk assessments, the study enabled the implementation of effective soil conservation measures and supported sustainable land management practices. Continued monitoring and the integration of advanced technologies will further enhance efforts to combat soil erosion, ensuring long-term agricultural productivity and environmental sustainability in the region.

6.8 Conclusion

Remote sensing, integrated with Geographic Information Systems (GIS) and other technologies, plays a pivotal role in watershed management by offering precise and comprehensive data for monitoring, assessment, and planning. In India, where diverse climatic conditions and varied topographies create unique challenges for watershed management, the use of these technologies has proven particularly beneficial. Here are detailed explanations with specific Indian examples to illustrate the applications and benefits:

1. Land Use and Land Cover Mapping

Example: Western Ghats

The Western Ghats, a UNESCO World Heritage site, is a biodiversity hotspot facing significant pressure from agricultural expansion, urbanization, and deforestation. Remote sensing data from satellites like Landsat and Sentinel-2 are used to monitor land use and land cover changes over time.

Implementation

- **Data Collection:** High-resolution satellite imagery is collected periodically.
- **Analysis:** Classification algorithms such as supervised classification and machine learning techniques are applied to differentiate between forest, agricultural land, urban areas, and water bodies.
- **GIS Integration:** The classified data is integrated into GIS to create detailed maps showing land cover changes over time.

Impact

- **Conservation Planning:** Helps in identifying areas that need protection and restoration.
- **Policy Making:** Informs policies to control deforestation and manage urban sprawl.
- **Biodiversity Management:** Supports conservation efforts by monitoring habitat changes.

2. Hydrological Modeling

Example: Narmada River Basin

The Narmada River Basin is critical for water supply, agriculture, and hydropower. Remote sensing data is used to derive inputs for hydrological models to predict water availability and flood risks.

Implementation

- **Data Collection:** Satellite data (e.g., precipitation data from TRMM/GPM, soil moisture from SMAP).
- **Analysis:** Remote sensing data is processed to derive hydrological parameters like precipitation, evapotranspiration, and soil moisture.
- **GIS Integration:** The data is integrated into hydrological models (e.g., SWAT, HEC-HMS) within a GIS framework to simulate river flow and predict flood events.

Impact

- **Flood Management:** Improved flood forecasting and risk assessment.
- **Water Resource Planning:** Better planning for irrigation, drinking water supply, and hydropower generation.
- **Disaster Preparedness:** Enhanced preparedness and response strategies for flood events.

3. Soil Erosion and Sedimentation

Example: Chambal River Basin

The Chambal River Basin is notorious for severe soil erosion, especially in its ravine regions. Remote sensing and GIS are used to map soil erosion patterns and implement soil conservation measures.

Implementation

- **Data Collection:** High-resolution LiDAR data and satellite imagery.
- **Analysis:** Digital Elevation Models (DEMs) created from LiDAR data to identify erosion features. Erosion risk maps generated using GIS.
- **GIS Integration:** Soil type, land use, and topographic data integrated into GIS for comprehensive analysis.

Impact

- **Erosion Control Measures:** Informed the design of terraces, check dams, and reforestation projects.
- **Policy Making:** Data-driven policies for soil conservation and land management.
- **Sustainable Agriculture:** Promotion of sustainable agricultural practices to reduce soil erosion.

4. Water Quality Monitoring

Example: Ganges River

The Ganges River suffers from pollution due to industrial discharge, agricultural runoff, and urban sewage. Remote sensing is used to monitor water quality and identify pollution sources.

Implementation

- **Data Collection:** Multispectral satellite data from MODIS and Sentinel-2.
- **Analysis:** Algorithms applied to estimate water quality parameters like chlorophyll concentration, turbidity, and suspended sediment load.

- **GIS Integration:** Water quality data integrated into GIS to create pollution maps and identify critical pollution sources.

Impact

- **Pollution Control:** Informed the implementation of pollution control measures and wastewater treatment facilities.
- **Public Health:** Improved water quality, contributing to better public health outcomes.
- **Regulatory Measures:** Strengthened regulatory frameworks for industrial and urban waste management.

5. Vegetation and Forest Management

Example: Sundarbans Mangrove Forest

The Sundarbans mangrove forest is crucial for coastal protection and biodiversity. Remote sensing helps monitor vegetation health and forest cover changes.

Implementation

- **Data Collection:** Satellite imagery from Landsat, Sentinel-2, and LiDAR data for detailed topographic information.
- **Analysis:** NDVI and other vegetation indices used to assess vegetation health. Change detection techniques applied to monitor forest cover changes.
- **GIS Integration:** Vegetation data integrated into GIS for spatial analysis and mapping.

Impact

- **Conservation Efforts:** Informed reforestation and mangrove restoration projects.
- **Biodiversity Protection:** Supported efforts to protect wildlife habitats and biodiversity.
- **Climate Change Adaptation:** Enhanced resilience against coastal erosion and sea-level rise.

Conclusion

The integration of remote sensing with GIS and other technologies provides a powerful toolkit for watershed management in India. By offering precise and comprehensive data, these technologies enable better monitoring, assessment, and planning, ensuring sustainable management of watershed resources. This, in turn, promotes environmental health, enhances community well-being, and supports economic development. The detailed examples from different regions of India illustrate the diverse applications and significant benefits of these technologies in addressing the country's unique watershed management challenges.

Important Questions

1. How has land cover in the Amazon Basin changed over the past decade, and what are the primary drivers of these changes according to Landsat imagery analysis?
2. Can you discuss the effectiveness of conservation efforts in the Amazon Basin based on the trends identified through land cover change detection using Landsat imagery?
3. How can remote sensing technologies, such as satellite imagery and LiDAR, be used to monitor vegetation health and assess forest cover changes over time?
4. What are some effective strategies for sustainable forest management, and how can remote sensing data be integrated into decision-making processes to support these strategies?
5. How has water quality in the Ganges River changed over recent years, and what are the primary factors contributing to these changes according to water quality monitoring data?
6. What are the key challenges in maintaining and improving water quality in the Ganges River, and how can the findings from water quality monitoring efforts inform effective management and policy decisions?
7. How do hydrological models, such as SWAT or HEC-HMS, simulate water flow and predict flood events in the Narmada River Basin, and what are the key inputs required for accurate modeling?
8. Can you discuss the role of hydrological modeling in water resource planning and management in the Narmada River Basin, and how have the findings from these models informed decision-making processes regarding irrigation, hydropower generation, and flood risk mitigation?
9. How does remote sensing technology contribute to the assessment of land cover changes and vegetation dynamics within watershed boundaries, and what are the key satellite sensors used for such analyses?
10. Can you elaborate on the integration of remote sensing data with Geographic Information Systems (GIS) for spatial analysis and mapping in watershed management, and how does this integration enhance decision-making processes?

11. What role does remote sensing play in monitoring hydrological processes such as precipitation patterns, soil moisture levels, and river flow dynamics within watersheds, and how does this information contribute to water resource management?
12. How do remote sensing techniques assist in identifying erosion-prone areas, assessing soil erosion rates, and implementing soil conservation measures within watersheds, and what are some examples of successful applications in real-world scenarios?
13. Can you discuss the role of remote sensing in water quality monitoring within watersheds, including the estimation of parameters such as turbidity, chlorophyll concentration, and pollutant levels, and how are these data utilized for environmental assessment and management?
14. How do multispectral and hyperspectral imaging differ in their applications and capabilities in environmental monitoring, and what are some examples of their use in detecting specific features or phenomena?
15. Can you discuss the advantages and limitations of LiDAR (Light Detection and Ranging) technology compared to traditional aerial photography for terrain mapping and vegetation structure analysis?
16. What role does Synthetic Aperture Radar (SAR) play in overcoming limitations posed by cloud cover and vegetation obstruction in satellite imagery, and how is SAR data utilized in applications such as flood monitoring and land deformation analysis?
17. How are advancements in satellite technology, such as improved resolution and the development of new sensors, enhancing the capabilities of remote sensing for monitoring environmental changes on both global and local scales, and what are the implications of these advancements for environmental management and policy-making?
18. What are some of the key challenges faced when using remote sensing technology in watershed management, particularly in relation to data resolution, accuracy, and the interpretation of imagery for diverse environmental contexts?
19. How can the limitations of remote sensing, such as the inability to penetrate dense vegetation or accurately assess subsurface features, be addressed or mitigated through the integration of ground-based monitoring techniques and modeling approaches in watershed management practices?
20. How does the use of Synthetic Aperture Radar (SAR) data contribute to improving flood risk assessment and management in the Godavari River

Basin, particularly in terms of its ability to provide real-time monitoring and accurate mapping of flood extents?

21. Can you discuss the challenges and limitations associated with utilizing SAR data for flood risk assessment in the Godavari River Basin, including issues related to data processing, interpretation, and the integration of SAR-derived information with other hydrological datasets?
22. How does the integration of remote sensing data with Geographic Information Systems (GIS) enhance the effectiveness of watershed management practices, and what specific benefits does this integration offer in terms of data analysis, visualization, and decision-making?
23. Can you elaborate on the challenges and considerations involved in integrating remote sensing data with other technologies, such as IoT sensors and hydrological models, within a GIS framework for comprehensive watershed management, and how are these challenges addressed in practice?
24. How does remote sensing technology, particularly multispectral and hyperspectral imaging, contribute to the mapping and monitoring of wetlands, and what are some key indicators used to assess wetland health and dynamics?
25. Can you discuss the challenges associated with wetland mapping and monitoring using remote sensing techniques, including issues related to data resolution, seasonal variations, and the classification of complex wetland ecosystems, and how are these challenges addressed in practice?

Index

D

E

T